AMERICAN AVIATION ICONS

NORTH AMERICAN AVIATION

F-100 SUPER SABRE

THEODORE GIANNA

ISBN: -------------------------- (Paperback)
First printing edition 2022.

CONTRIBUTORS AND RESOURCES:

I want to thank a few people who helped publish this book.

First, I want to thank someone I am sure many aviation enthusiasts have heard of: Joe Bougher. He allowed me to use his vast resources, and I am grateful. Take a look through Joe's informative website at: http://www.joebaugher.com/

I would also like to thank Michael Fader and Eric Sanvoisin. They are two of the most talented aircraft profile artists that I have come across. Their excellent work is on full display in this book, and I warmly thank them for their contributions. I would also urge the readers to check out Michael's website. I am sure you would find it very interesting: http://www.wings-aviation.ch/

Most of the photos in this book are from the US Air Force, the National Museum of the US Air Force in Dayton, Ohio, the world's largest military aviation museum, and NASA.

INTRODUCTION

This book is a history of the F-100 Super Sabre. The F-100 kicks off the series because it broke many barriers. On October 29, 1953, the YF-100A prototype piloted by NAA Chief test pilot George Welch set a world speed record of 755.149 mph in level flight. Then, on August 20, 1955, an F-100C set a supersonic world speed record of 822.135 mph. The F-100 was the first small- to medium-sized jet to carry an atomic weapon, the MK 7 nuclear bomb. It became the first aircraft to successfully perform buddy refueling when two F-100Ds performed the maneuver on December 26, 1956. It was also the first of a particular breed: the famous Century Series of jet fighters. It was an icon in the early 1950s and had its day in the sun. Later, more advanced aircraft overtook it. By 1960, the F-100 was considered too old. However, the Vietnam War was a godsend for the F-100. The Super Sabre finally found its calling as a close-support fighter-bomber during the volatile Vietnam conflict. It flew more sorties than any other aircraft during its Southeast Asia service. The 'Hun,' as it was known, served the USAF well until the Air Force retired it in 1971.

NORTH AMERICAN F-100 SUPER SABRE

Development of the North American Aviation (NAA) F-100 Super-Sabre, the first of the 'Century Series' of USAF fighters, began on February 3, 1949. After the success of the F-86 Sabre program, NAA began a new project to create its first jet aircraft capable of breaking the sound barrier in level flight. The company funded the project in-house, led by two great designers, Ray Rice and Ed Schmued, who created the famed P-51 Mustang. If anyone could overcome the obstacles of creating a new aircraft, especially one that could break the sound barrier, it was Rice and Schmued.

The YF-86 that NAA used for tests on the Sabre 45 project. (NASA)

From the start, the designers knew the company's first supersonic fighter couldn't be a Sabre clone. The new aircraft would share little with the F-86. Its engine needed to be significantly more powerful than the Sabre's, which meant a larger engine and a complete airframe redesign, resulting in a larger, stronger aircraft with much higher performance than the F-86. It also meant the new jet would have the power, size, and weight to push through the sound barrier by brute force.

As part of the design development, NAA commissioned a supersonic tunnel, one of the first supersonic tunnels built by the private sector. As a result, on September 14, 1949, NAA began wind tunnel testing on an F-86D with the new 45-degree swept wing. The modified F-86D was commonly referred to as the 'Sabre 45' at NAA because of the aircraft's forty-five-degree wing sweepback compared to the F-86's 35-degree.

The tests led to a modified F-86D with a swept wing increased to 45 degrees, a tailplane, and a contoured fuselage to conform to the 'area rule' (the principle used to design aircraft with minimal drag at speeds approaching and exceeding Mach 1). The NAA test plane was estimated to have 50% less supersonic drag than other jet fighters at the time.

The final design was acceptable; however, the aircraft suffered from a few imperfections, some of which were controversial. For example, the wings had significant inboard ailerons to ensure good lateral control and a high roll rate without excessive wing twist and subsequent control reversal. However, that meant that no flaps could be fitted. No flaps on a thin supersonic wing, such as the F-100 had, was a problem. It would result in an even higher landing speed than such a wing required. The most NAA could do was install a massive full-span slat along the leading edge. That didn't sit well with some pilots, who complained that landing the F-100 felt like a controlled crash.

The horizontal tailplane was a slab with no elevators, but it moved fully. It sat at the aircraft's lowest point, in the wake of the wings, helping prevent pitch-up. To further assist stall control, most F-100 models had wing fences. The F-100 also had an enormous single-speed brake: an awkward-looking metal sheet that resembled a large elevator door with a cutout section in the lower middle. It was situated in the aircraft's underbelly, near the center of gravity.

An F-100A (52-5778) Super Sabre on the Rogers Dry Lakebed, photographed in 1955. The NACA High-Speed Flight Station at Edwards, California, operated this F-100 Super Sabre from 1954 to 1960 to investigate the stability and control characteristics of the then-new supersonic Air Force fighter. (NASA)

YF-100A 52-5754 was the first to break the sound barrier in sustained level flight. It is seen here flying over Edward's AFB. (USAF)

YF-100A 52-5754. On display at Petersen AFB in 1963. Scrapped 1968. (USAF)

The brake would blow back furiously against the airframe at extremely high airspeeds. Who said designing planes was easy?

Interestingly, the F-100 Super Sabre got a nickname: the 'Hun.' Not after the nomadic warlike Asiatic tribes that invaded and ravaged Europe in the 4th to 5th centuries. It was short for 'Hundred.' One other nickname was reserved for the RF-100A reconnaissance version: 'Slick Chick.'

After over two years of collaboration with the Air Force, North American completed its (NA-180) project on January 19, 1951.

The Air Force's top leadership was preoccupied with the Korean War. In the early stages of the war, in June 1950, the Air Force was doing well. The USAF Far East Air Forces flew piston-engine F-51D Mustangs, all-weather F-82 Twin Mustangs, and jet-powered F-80 Shooting Stars. Skilled USAF pilots soundly defeated the inexperienced North Korean Air Force (NKAF), which flew around 140 World War II-era piston-engine aircraft. After defeating the NKAF, the USAF and the other UN air forces dominated the skies over Korea until the introduction of the MiG-15 in November 1950. The MiG-15, flown by skilled Soviet pilots, threatened to take control of the air from USAF and UN forces. The Mig-15 even outgunned and outmaneuvered the strongest USAF fighter in Korea, the F-80C. Then, in mid-December, the USAF responded to the MiG menace with the F-86 Saber

The YF100A 52-5754 with a break chute (USAF)

One of the first F-100A Huns, F-100A-1-NA 52-5761. Now on display at the New England Air Museum. (USAF)

F-100A 53-1530 in flight over mountains. To Republic of China AF as 31530, WFU. (USAF)

jet fighter. The F-86As entered combat and quickly proved their effectiveness by reclaiming the skies. The Soviets' brief control of the skies over Korea worried USAF officials. After the impact of the first MiG-15s in the Korean conflict, the USAF expressed interest in the F-100 and issued a General Operational Requirement (GOR) on August 27, 1951. The Air Force requested a fighter for the air superiority role, similar to the 'Sabre 45.' The MiG-15's arrival in the Korean conflict compelled the Air Force Council and the Aircraft & Weapons Board to regard the GOR as a response to a national emergency to provide funding for the 'Sabre 45' production batches, even before the prototype had been evaluated. The USAF needed new fighters, and the F-100 was top of the list. On November 1, 1951, the USAF ordered two YF-100 prototypes, later redesignated YF-100A, and then ordered 110 F-100As. Then, on November 7, 1951, USAF representatives viewed the aircraft's scale model and requested over a hundred modifications.

On August 26, 1952, the USAF approved the F-100A mock-up, which included many changes. The F-100A retained two of the F-86's best features: good forward cockpit visibility and a clear, unobstructed forward nose intake. However, it underwent some modifications. To improve cockpit visibility, the Hun's nose intake was lengthened laterally and flattened, creating the typical gaping snout with sharp lips in an almost fishlike mouth shape. The front intake's steep lip was intended to be efficient at near-sonic speeds, but it inevitably caused turbulence and power loss at lower speeds. The unusually flat fuselage bottom was presumably designed to mitigate any adverse effects when landing.

The F-100's wings had a 6% thickness ratio. As mentioned earlier, the wings also had slats and wide inboard ailerons instead of flaps. With wide angles of sweep, conventional wingtip ailerons imposed substantial twisting loads at high speeds, inducing control reversal and becoming ineffective at very high and very low speeds because airflow spills from the tips. In contrast, when inboard ailerons are attached where the wing is thicker and stiffer, the ailerons' leverage around the root is reduced.

The cockpit cover was enlarged and designed to provide the pilot with an excellent overall view of his surroundings. The sloping windshield was clearly designed for speed, but some pilots disliked that shape. However, inside the cockpit, the top of the F-100A's ejector seat and the armor did not restrict the pilot's rear view, as in many other contemporary fighters of the 1950s.

Finally, the spine running from the cockpit cover to the fin suppressed airflow turbulence behind the cockpit. However, that could cause the boundary layer to break away on the fin, rendering much of that area ineffective at near-sonic speed.

F-100A

561st Tactical Fighter – North American F-100A-5-NA Super Sabre 52-5777. On display at the Hill AFB Museum, Utah, since 1990. (USAF)

The prototype YF-100A was completed on April 24, 1953, and George Welch, NAA's chief test pilot, flew it for the first time a month later. Welch flew the YF-100A, 52-5754, powered by a Pratt & Whitney XJ57-P-7 engine, at Edwards AFB on May 25, 1953, and broke the sound barrier. He followed it up with the second YF-100A, 52-5755, on October 14, 1953. The Air Force completed Phase 2 flight tests on September 17, 1953, and on October 29, the prototype Super-Sabre set a world speed record at sea level of 755.149 mph, piloted by Lt.-Col. Pete Everest. On the same day, George Welch flew the first production F-100A, 52-5756. Coincidentally, the Super Sabre prototype was the 50,000th aircraft North American produced since its founding.

The remarkable YF-100A speed feats would not have been possible without a powerplant. That powerplant was the Pratt & Whitney J57 engine with an afterburner. It was the F-100's engine of choice. The Boeing XB-52 was the first to use it. In the early 1950s, the J57 was considered the most powerful and successful of any turbojet engine produced in the United States. It was an axial-flow turbojet engine. Its two-spool or dual-compressor design achieved previously unattainable pressure ratios, significantly reducing fuel consumption. It widened the engine's efficient operating range. Furthermore, the J57 was the first US-manufactured engine to attain 10,000 lbf (45 kN) thrust.

On October 20, 1953, five months after the prototype's maiden flight, the main Los Angeles factory began regular production. The company expanded production to the Columbus (Ohio) plant on September 19, 1954, after the USAF placed a second large order. The first F-100As, still unmodified (their initial operational capability was declared valid only in September 1955), entered service with the 479th Day Fighter Wing at George AFB, based in Victorville, California, in September 1954. The F-100As entered service immediately. While this may have seemed like a rapid transition, it was not. Air Force pilots who would form the first American supersonic squadrons took part in formal tests in early 1954. As a result, the pilots were all ready to go when the first batch of F-100As arrived in September.

F-100A of the 152nd Fighter Interceptor Squadron, 53-1639, To MASDC, May 11, 1966. Then to the Republic of China (Taiwan) AF as 31639, code 0233. Now on display at Hualien, Taiwan. (USAF)

F-100A 52-5760 going vertical. This aircraft was redesignated JF-100A on Nov. 28, 1955. Then, on Feb. 28, 1960, it was redesignated again as NF-100A. On Sep. 10, 1971, it became a Ground Instruction aircraft with code TJ. On Aug. 21, 1972, it deteriorated abnormally and was stored at Edwards AFB, California, until the Fall of 2003. (USAF)

All was going according to plan at NAA. Development of the F-100 was remarkably trouble-free, and production at the NAA Inglewood, California, facility, which during the war manufactured the famed P-51 Mustang and B-25 Mitchell bomber, ramped up impressively. Then tragedy struck.

On October 12, 1954, George Welch was assigned to test the aircraft's structural limits by performing a series of extreme maneuvers. One was a high-speed, violent 7.8 G turn. As agreed, Welch flew his plane through several maneuvers before attempting the 7.8 G maneuver. First, he put the aircraft nose-down to reach Mach 1.5. He then executed a diving left roll at 25,000 feet. After Welch performed the maneuver, the F-100A went wildly out of control as its vertical stabilizer broke off and the jet disintegrated in midair. The crew of a B-47 bomber flying near the F-100 test zone saw debris fall from the sky, including two parachutes. The first chute was the drag chute, violently ripped out of its container. The second chute was drifting slowly towards the ground with George Welch's mortally wounded body. A courageous young pilot died that tragic day. But let's not forget that he was a World War II flying ace, a Medal of Honor nominee, and an excellent test pilot.

Kenneth M. Taylor and George Welch, shortly after the Pearl Harbor attack. During the attack, both men took two Curtiss-Wright P-40B Warhawks into the air and, against overwhelming odds, each shot down four enemy aircraft. (USAF)

Six additional F-100As crashed afterward under identical circumstances and for similarly inexplicable reasons. As a result, the Air Force grounded all F-100s on November 10, 1954. The Air Force, NAA, and NACA generated complementary research programs to analyze and fix the troubling issue.

The reason for the accidents was discovered after a lengthy and complex investigation. The accidents stemmed from the prototype F-100A Super Saber's initial design: sharp swept wings, a long fuselage, and a small vertical fin, a recipe for disaster. While other fighter pilots were in awe of the F-100's speed, the USAF F-100A pilots who flew it were far less confident. They noticed that the Hun's lateral directional stability was unsatisfactory throughout the entire combat speed range, with lateral directional oscillations showing no tendency to dampen. Even after this damning proclamation by the USAF pilots, North American shrank the F-100A's vertical fin further instead of increasing it. By lowering the fin's height, NAA reduced its area and aspect ratio and increased its taper ratio. Unfortunately, these changes did not help. In fact, they further reduced the F-100A's directional stability, by some estimates, by as much as 50% compared with the YF-100. The first production F-100As entered service in the late summer of 1954. During this period, inertial coupling came to the fore, killing the effervescent George Welch. The accidents immediately grounded all the F-100s. At the same time, the Air Force, North American Aviation, and the National Advisory Committee for Aeronautics (NACA) crafted complementary research procedures to analyze and fix the troubled program. The cause of the accidents, which took several pilots' lives, was inertial coupling.

This is an RF-100A 'Slick Chick', 53-1551. It was originally an F-100A and was modified in 1954 into an RF-100A unarmed photographic reconnaissance aircraft. This aircraft served with Detachment 1 of the 7407th Support Squadron, 7499th Support Group. It crashed near Neidenbach, Germany, on October 1, 1956. The pilot ejected safely. (USAF)

Inertial coupling arises when an aircraft's flight speed progresses from subsonic to transonic and into the supersonic. These transitions change the proportional relationship between the wings and the fuselage. As the aircraft's speed rises, the span-to-fuselage length ratio decreases. As a result, the airplane's mass distribution stretches along the fuselage. However, on the small, low-aspect-ratio wings of supersonic jets like the F-100A, roll stability drops dramatically, to the point of being virtually nonexistent. This situation would cause a violent, uncontrollable yawing motion. It would be like being on an extremely wild bucking horse at the rodeo. This ferocious yawing action precipitated the crash of Welch's F-100A and took his life. Thus, any high-speed roll or turn, especially the 7.8 G left roll turn by Welch's F-100A, would generate exceptionally large sideslip loads on the aircraft's vertical fin, causing structural failure, the fin breaking off and the plane disintegrating.

The Air Force and NACA were concerned about the F-100A since it was the first of an entirely new class of supersonic fighter aircraft, the coveted 'Century Series' that the USAF, and at least some of its allies, intended to reequip their fleets with. Solving the F-100's inertial-coupling problems was urgent. However, it took months of extensive NACA and Air Force flight testing, much of it very high risk, before they knew the cause of the crashes. In the end, the F-100A needed a significant redesign. The multiple Super Sabre accidents underlined that streamlined supersonic aircraft needed a significantly larger vertical stabilizer area, along with artificial stability and motion damping, to keep sideslip from developing to a dangerous, uncontrollable level. As a result, all F-100s in production and under construction required modifications. The F-100A fin surface was expanded by 27%, and its height increased by 15 cm. The F-100A's wingspan was stretched by 60 cm at the wingtips, and the surface controls were strengthened. When the F-100A returned to service in February 1955, not all its problems were fixed, notably its unpredictable engine, which had suffered several failures. In late 1955, the last F-100A was delivered. By then, total production had reached 203 units. The final F-100As to leave the factory were better than the earlier F-100As. They benefited from improvements, including a redesigned cockpit layout, updated instruments, and, more importantly, a new, more powerful J57-P-39 engine.

In the end, the initial version of the Super Sabre, the F-100A, did not prove to be very advantageous. It had limited offensive capabilities, and its equipment was not very sophisticated. It was never intended to be an all-weather fighter and mainly served as a day fighter. In fact, it was superior to the F-86 only in speed and power, thanks to its Pratt & Whitney J57-P-7/39 engines. However, the F-100A had a rudimentary Identification Friend or Foe (IFF) unit. As for radar, it had only a radar gunsight ranging system, the AN/APG-30, used to aim the four Pontiac-produced M-39 20mm cannons. It had no separate radar display, and the pilot had to acquire the target visually. It was the only offensive radar system the F-100 ever carried. Its four 20 mm M-39E cannons were installed under the cockpit, with 200 rounds per weapon. In Korea, the M-39 was tested on a modified F-86F and had a rate of fire of about 1,500 rounds per minute, with an initial shell speed of 625 mph. Interestingly, the M-39 was mounted on the F-86H and later on the McDonnell F-101A. By the end of 1955, the USAF assigned all F-100As to tactical squadrons as day fighters until the more versatile F-100C replaced them in 1957.

The USAF examined the potential of converting some F-100As into reconnaissance variants. As a result, the USAF converted six F-100As to RF-100A configurations and, as mentioned earlier, nicknamed them 'Slick Chicks'; they flew with no armament. Instead, they carried camera equipment in an extended section of the forward lower fuselage. They also

On the left is F-100A, 52-5778, with a modified vertical fin. On the right is F-100A, 52-5773, with the original smaller vertical fin configuration. The NACA added a larger vertical fin to the airplane in December 1954, increasing its surface area by 10%. Later, NAA installed an even larger fin, increasing the area by 27%. NAA also extended the aircraft's wingtips. These modifications solved the dangerous directional stability and roll-coupling problems the F-100A was experiencing. (USAF)

carried four external fuel tanks. After flying with the USAF in Europe (USAFE) and the Far East Air Forces, the USAF delivered four of these aircraft to Taiwan (then known as Nationalist China) in 1961.

NAA's next model should have been the F-100B. That was not the case. In October 1953, the all-weather F-100B derivative was designed and later became the NA-212. The concept eventually evolved into the F-107 fighter-bomber, powered by a Pratt & Whitney J75 engine producing 23,500 lb. of thrust. The F-107 design was one-of-a-kind because its air inlet sat above and behind the cockpit. The complex aircraft was still in development in 1955. As a result, NAA switched its assembly lines from the F-100A to the new F-100C ground-attack variant.

F-100C

As the USAF fighter-bomber fleet aged and relied mostly on Republic F-84s, North American offered its new NA-214 model (later F-100C) to the Air Force; the Air Force approved it on December 30, 1953, and placed a production contract in February 1954.

In 1955, just over a year after USAF approval, the first F-100C flight, 53-1709, took off from Los Angeles on January 17, 1955, with Al White at the controls. They chose Los Angeles for the event because North American Aviation manufactured the aircraft at its Downey Plant there. In fact, NAA's Los Angeles plant built 451 F-100Cs, and its Columbus, Ohio, plant built another 25, bringing total F-100C production to 476.

The F-100C had a reinforced wing, which the F-100A-1 (52-5759) tested in July 1954. The F-100C could carry much more fuel than the F-100A. It had a 'wet wing,' meaning the fuel was no longer stored only in separate tanks on the aircraft's wings but also within the wing structure. It also installed fuselage fuel tanks. As a result, internal fuel capacity increased from 744 gallons in the F-100A to 1,702 gallons in the F-100C. Two 275-gallon tanks, augmented by two 200-gallon tanks, carried external fuel. The extra 200-gallon tanks extended range. However, when these tanks were used with the 275-gallon tanks on the inboard stations, they reduced directional stability, which was especially critical at speeds above Mach 0.8. This problem was not fixed because of the high costs involved. Nonetheless, the Air Force came up with a solution. Tests revealed that larger external tanks did not compromise the F-100C's longitudinal stability. The Air Force fixed the problem by banning the 200-gallon tanks and ordering 450-gallon external tanks instead of the 275- and 200-gallon combination. Carrying more fuel than the F-100A, the F-100C had a greater range while allowing a higher weapons load.

F-100C-20-NA Super Sabre, 54-1894. 136 TFS Niagara Falls ANG, NY. On March 4, 1968, during preparations for SEA deployment, the pilot was killed in midair collision with 54-1900 over Lake Ontario, whilst on an air combat maneuvering mission. (USAF)

While the F-100A could carry only 1,000 lb. of bombs, the F-100C had six wing hardpoints for bombs or other munitions totaling up to 5,000 lb. Regarding air-to-air combat, the F-100C was not much better than the F-100A in armament. The F-100C had the usual four 20 mm Pontiac M-39 guns and the same APG-30 gunnery radar as the F-100A. However, with a higher weapons-carriage capability than the F-100A series, the F-100C could carry a tactical nuclear bomb, the Mk 7, thanks to a new delivery technique: the Low Altitude Bombing System (LABS). This technique would enable the F-100C to release the Mk 7 nuclear bomb at a low altitude and swiftly escape the shock waves. According to this system, the aircraft would approach the target at low altitude and high speed. Shortly before flying over the target, it initiates a pitch-up, during which it releases the nuclear weapon. While the bomb follows a ballistic trajectory that brings it down onto the target, the aircraft continues to the top of the loop. Then, instead of completing the loop, it straightens with a half roll and shoots away at high altitude. The F-100C was an ideal platform for the low-altitude bombing system (LABS) since its maximum ground speed of 1,050 feet per second was significantly greater than the delivery speed of operational fighters at the time.

The F-100C used the J57-P-7-21 engine, with 10,200 lb. military thrust and 16,000 lb. afterburner. This engine raised the F-100C's top speed to Mach 1.25 at low altitude, despite adding nearly 2,600 lb. of combat weight. It also gave the fighter-bomber an extra 600 miles of range. With the new J57-P-21 engine, Colonel Horace A. Haines set new speed records on August 20, 1955, in an F-100 at about 40,000 feet. The tests took place over the Mojave Desert on an 11-mile circuit. Col. Haines recorded speeds of 870.627 mph in one direction and 733.644 mph in the opposite direction. In the two-stage test, he averaged 822.135 mph. However, official statistics show that the F-100C's peak speed exceeded Col. Haines's. The F-100C's maximum speed might have reached 924 mph at 35,000 feet and 904 mph at 39,500 feet. Subsequently, Colonel Horace A. Haines' record was a conservative estimate of the F-100C's actual top speed.

The F-100C also had another problem that NAA fixed in 1955. In its primary day-fighter role, the F-100C, like the F-100A, tended to yaw and eventually enter an uncontrolled roll at high speeds. NAA installed a hydraulically actuated, electrically operated yaw damper, starting with the 146th F-100C production aircraft, which substantially improved performance. Consequently, with NAA assistance, the Air Force modified the first 145 F-100Cs with the device. Similarly, installing a pitch damper in the horizontal stabilizer control system boosted the damping of longitudinal oscillations. Factory installation of the pitch damper was initiated on the 301st F-100C.

The Air Force proposed installing the detachable in-flight refueling probe beneath the F-100C's right wing. With this design modification, the F-100C could reach inaccessible regions in just a few hours. In addition, the fact that the F-100C could transport a Mark 7 nuclear warhead made it a very dangerous aircraft. As a result, the Tactical Air Command (TAC) of the United States Air Force now had its first supersonic atomic bomber.

The first F-100Cs with an in-flight refueling probe arrived at the 450th Day Fighter Wing on July 14, 1955, based at Foster AFB, Texas (*by the end of 1956, the 450th DFW became the 322nd Day Fighter Group). While other TAC units took delivery of their F-100s, the USAF Aerobatic Team, then called the 'Skyblazers,' chose the F-100C Super Sabre, whose impressive performance helped the team's flight demonstrations succeed.

The F-100C's inability to carry all the USAF's primary non-nuclear weapons was a thorn in the Air Force's side. The F-100C's weapons-carrying capacity was finally enhanced in 1962, seven years after the first F-100C was introduced! It replaced the two previously installed hardpoints with six MA-3-type ejection hardpoints. In addition, the more advanced AN/ASC-17 radar replaced the obsolete AB/AP6-30A type firing radar. Even so, the F-100C's ability to carry and fire conventional armaments was severely constrained. By the end of 1965, only 125 ANG F-100Cs could drop anti-personnel CBU cluster bombs and carry AIM-9B Sidewinder air-to-air missiles.

Furthermore, the F-100 fuel tank problem, first identified in 1955 (see above), persisted in 1962. The USAF proposed installing 450-gallon tanks to fix this. However, these tanks were hard to obtain and costly. The TAC then recommended replacing the 450-gallon drums with 355-gallon drums. The Air Force approved the project, but many technical difficulties nevertheless remained.

121st Tactical Fighter Squadron F-100C-1-NA 53-1727. Notice the refueling probe under the right wing. To Turkish AF, November 20, 1974, as 53-1727 (USAF)

An F-100C-25-NA. This F-100 was involved in an accident on Nov. 27, 1959, at Luke AFB, Arizona. (USAF)

JF-100C

The F-100C was retitled JF-100C, assigned serial number 53-1709, and sent to the NACA High-Speed Flight Station at Edwards AFB, California. NACA flew and tested this F-100 Super Sabre from 1954 to 1960 to investigate the stability and control features of the then-new supersonic Air Force fighter. On September 27, 1959, 'NACA High Speed Flight Station' was renamed 'NASA Ames Flight Research Center.' On November 2, 1960, 'NASA Ames Flight Research Center' transferred the JF-100C to 'NASA Langley Flight Research Center' at Langley Field, Virginia. At Langley Field, the JF-100C supported a range of airborne simulation studies, including some supporting the X-15 and SST studies. NASA sent the JF-100C to General Electric on February 3, 1964, to restore it to flight configuration. On March 11, 1964, it returned to NASA Ames Research Center in California, where it remained until May 21, 1972.

NASA re-registered the aircraft as #703 and transferred it to NASA Dryden Flight Research Center at Edwards AFB, California. While there, it acquired the tail surfaces of an F-100D. NASA retired the JF-100C in 1972 and donated it to San Jose State University, where it was used for maintenance training at the university's aviation school at SJC (Mineta San Jose International Airport). On August 2, 1974, its registration was canceled. Consequently, in 2004, the JF-100C was relocated to a temporary hangar at Castle Air Museum for repair. It is currently on display there and was assignedserial number55-2879, which was actually the serial number of an F -100D-50-NH.

The JF-100C, 53-1709, remained with NACA/NASA for most of its service (NASA).

TF-100C

A contract to produce a tandem trainer F-100 was placed in December 1955. The mock-up inspection was completed the following month using an F-100C-5-NA, 54-1966. The 'C' was converted into a two-seater version with a 36-inch extended fuselage for tandem cockpits beneath a single long canopy. The TF-100C first flew on August 6, 1956, and served as the prototype for the upcoming F-100F. The aircraft was fully equipped for combat missions but carried reduced armament: only two M-39 20 mm cannons, instead of the four standard cannons on other 'C' variants. This was required because the lengthened cockpit occupied the space used for the ammunition cans on the single-seat versions. Regrettably, just a few months later, on April 9, 1957, the JF-100C broke apart in midair northwest of Kramer Junction in the Mojave Desert. Fortunately, the pilot ejected safely.

The Thunderbirds were the first in the world to perform aerial demonstrations at supersonic speeds after switching to the F-100C Super Sabre in 1956. That same year, the Thunderbirds moved to their new home at Nellis Air Force Base in Nevada, which simplified aircraft maintenance and logistics. Between 1956 and 1963, the Thunderbirds flew the F-100C-model Super Sabre. In 1963, the supersonic aerial team used the F-105 Thunderchief. However, after safety concerns, they adopted the F-100D from 1964 to 1968.

In 1959, the Air Force started transferring the F-100Cs to the Air National Guard, where they were mobilized during the Berlin Crisis of 1961, when the East German government began building the infamous wall between East and West Berlin. Then again, during the Pueblo incident in 1968, when North Korea 'hijacked' the USS Pueblo in international waters. By June 1970, the Air Force had retired all F-100Cs. During their time with the Air Force and the ANG, the F-100Cs were involved in 85 serious accidents.

The TF-100C served as the prototype for the F-100F two-seat trainer. (USAF)

A North American F-100D drops a dummy nuclear bomb as it flies up to execute the Low-Altitude Bombing System (LABS) loop maneuver. (USAF)

F-100D

An F-100D Super Sabre from the USAF Thunderbirds aerobatic team at the National Museum of the United States Air Force at Dayton, Ohio. (USAF Museum)

On September 27, 1954, the Air Force designated an enhanced Super Sabre as the F-100D, and North American labeled it the NA-223. In October 1954, the Inglewood and Columbus factories completed a detailed study of the newest Super Sabre version. Then, in July 1955, the plants began aircraft production. The first F-100D-1-NA, 54-2121, which could exceed Mach 1.3, was rolled out in November 1955, and on January 24, 1956, Dan Darnell was the first pilot to fly the 'D' version.

The F-100D was the definitive F-100 variant. In addition, Los Angeles also manufactured 940 NA-223 and NA-235 versions with Sidewinder missiles, and Columbus produced another 334 NA-224s. Overall, North American built 1,274 F-100Ds.

The F-100D had many modifications, including a revised secondary wing construction with a curved trailing edge, broad inboard slotted flaps, two-piece ailerons placed farther out, and a higher fin. It also had a Minneapolis-Honeywell autopilot. The MB-3 autopilot provided stable control in supersonic flight, allowing the pilot to focus on the attack with either conventional or nuclear weapons, which could be carried out using the LABS procedure. The F-100D could carry not only the Mk.7 tactical fission bomb but also the Mk.38 and Mk.43 air-dropped variable-yield thermonuclear weapons. Alternate underwing loads included six 750-pound bombs, four 1,000-pound bombs, one 1,680-pound bomb, two 450-gallon air-refuelable buddy tanks, or four GAR-8 Sidewinder air-to-air missiles, and two GAM-83A Bullpup missiles. The Martin GAM-83A Bullpup ASM short-range air-to-ground missile was tested and fired from an F-100D-25-NA, 55-3681. The tests were completed in 1959. However, delays in missile delivery meant the first operational squadron did not receive its missiles until 1960, and three other squadrons received them in 1961.

Internal fuel storage increased significantly by sealing each wing's main bay as an integrated tank. The F-100D could also be equipped with special tanks that stored fuel for refueling between Super Sabers, or 'buddy tanking.' Other modifications included small fences above the outer wings, taller fins, and a deeper tank-vent fairing that eventually housed ECM and other aerials.

The F-100D was powered by the Pratt & Whitney J57-P-21A, which offered 10,200 lb. military thrust and 16,000 lb. with afterburner. The 'D' had a maximum speed of 864 mph.

The F-100D was the heaviest of the F-100 family, weighing 21,000 lb. empty and 34,832 lb. at maximum takeoff weight. However, compared to the 'C,' it had a lower top speed: 924 mph versus 864 mph for the 'F-100D.'

356th Tactical Fighter Squadron F-100D 56-3417 Aircraft now at 'Wings Over The Rockies Aviation and Space Museum', Colorado. (USAF)

494th Tactical Fighter Squadron F-100D-15-NA, 54-2277, in 1960. Sold to the Turkish AF on February 27, 1959, as 54-2277. (USAF)

352nd Tactical Fighter Squadron F-100D, 56-3373, in 1962. On Nov. 9, 1972, the jet's flight controls failed during a loop over Colorado. The pilot ejected safely (USAF).

A further increase in the internal tanks' capacity gave the F-100D a greater maximum range: 1,995 miles. However, its combat range of 534 miles was less than that of the F-100C's 572 miles. That wasn't an issue, though, because the F-100D was equipped with an in-flight refueling probe. Overall, the 'D' version was a daytime fighter-bomber with secondary air superiority capabilities.

Like its predecessors, the 'D' had its problems. Despite numerous attempts to improve the F-100D's autopilot, the process was delayed. The improved autopilot was scheduled to begin installation with the 184th F-100D production, along with the first production provisions for Sidewinder missiles and special centerline stores. However, it was not installed until the 384th 'D' production. Following that, all F-100Ds in service were retrofitted.

The small high-lift flaps between the ailerons and the fuselage improved low-speed handling qualities and were retained on the F-100D. But the 'D' still suffered from instability, especially when landing. Meanwhile, in early 1967, the F-100D's high-altitude maneuvering issues were finally resolved. The necessary modifications began with the aircraft's 226th manufacturing run. In addition, the older F-100Ds were backfitted. The F-100D also had air-refueling problems. During maneuvers under high 'G' forces, the in-flight refueling boom of many -F100Ds would detach from the wing. This prompted the removal of the refueling probes from all F-100Ds, except for air-refueling missions, until the aircraft's underwing structure could be reinforced.

Between 1962 and 1965, two significant changes were applied to the F-100D. First, as with earlier versions of the F-100 model series, the D's combat life was extended in 1962, as was its capacity to deploy non-nuclear weapons. The second change was even more critical and was called the High Wire project. The project was the most significant program carried out jointly by NAA and the Air Force. The project's primary objective was to standardize the F-100C and F-100D weapons systems.

An F-100D Super Sabre (s/n 55-2881) from the 531st Tactical Fighter Squadron, 3rd Tactical Fighter Wing, flying from Bien Hoa Air Base in South Vietnam. This aircraft was retired to the MASDC on April 18, 1979, designated FE0510. It was later converted to a QF-100D drone No.313. It was W/o on Nov. 14, 1988, when it was shot down. (USAF, Wikimedia)

401st TFW, 307th TFS F-100D, 55-3731, over Rogers Dry Lake in the Mojave Desert. This F-100D crashed in Spain on June 18, 1968. (USAF)

356th Tactical Fighter Squadron F-100D, 56-3418, seen here at Aviano AB, Italy, in 1960. It was involved in an accident at Cigli AB, Turkey, on May 1, 1969, and was W/O on May 22, 1969. (USAF)

An F-100D Super Sabre, s/n 55-2814, landing at Lakenheath, England, UK, on June 6, 1970. The aircraft was involved in an accident on Jan. 18, 1974, at Lambert, St. Louis, Missouri. (Steve Williams)

An F-100D-75-NA Super Sabre, s/n 56-3198, nicknamed "The Lemon", of the 615th Tactical Fighter Squadron, 35th Tactical Fighter Wing, being pushed into a revetment at Phan Rang Air Base, Vietnam, on June 13, 1970. (NARA – USAF photo code 342-C-KE-40616)

Three F-100Ds of the 481st TFS over Vietnam in 1966. Early F-100s arrived in Southeast Asia unpainted, like the foreground aircraft, but all eventually received SEA camouflage, like the aircraft in the back. From front to back, the aircraft are: F-100D-25-NA, 55-3647; F-100D-20-NA, 55-3559; and F-100D-20-NA, 55-3572. (USAF)

An F-100D s/n 56-3238, of the 50th Tactical Fighter Wing (Wing Commander's Aircraft), at Toul Air Base, France, in 1958.

F-100D-15-NA, 54-2281, was delivered to the USAF in 1956. In 1973, it was mounted on a pylon at Harry Bonsall Sr. Park at Glendale, Arizona. (USAF)

F-100D-75-NA, 56-3150, of the 522nd TFS. Later, with the 428th TFS, 474th TFW. The aircraft was hit by AAA while attacking Quang Khe Naval Base, at the mouth of the Sou Giang River in North Vietnam, on March 2, 1965. The pilot ejected over the Gulf of Tonkin and was rescued by a USAF HU-16 Albatross. (USAF)

F-100D-90-NA Super Sabre (56-3307) USAFE 417th TFS, 50th TFW, BA 136 Toul-Rosières, France 1959. To MASDC, Davis Monthan AFB February, 1979 as #FE463.Converted to QF-100D drone program #292.

F-100D-60-NA Super Sabre (56-2933) 1st FDS, 413th FDW, George AFB, California 1959.

F-100D-65-NA Super Sabre (56-3000) USAFE 77th TFS, 20th TFW, RAF Wethersfield, Essex UK 1958. On display outside at Texas ANG, Lackland AFB, Kelly Field, San Antonio, Texas

F-100D-80-NH Super Sabre (56-3357) 352nd TFS, 354th TFW, Myrtle Beach AFB, South Carolina 1957.

F-100D-90-NA Super Sabre (56-3307) USAFE 417th TFS, 50th TFW, BA 136 Toul-Rosières, France 1959. To MASDC, Davis Monthan AFB February, 1979 as #FE463.Converted to QF-100D drone program #292.

F-100D-60-NA Super Sabre (56-2933) 1st FDS, 413th FDW, George AFB, California 1959.

F-100D-70-NA Super Sabre (56-3101) Squadron Commander's Aircraft. 429th TFS, 474th TFW, Cannon AFB, New Mexico, 1958. To MASDC, Davis-Monthan AFB, July, 1978 as #FE392. Converted to QF-100D drone program #340.
F-100D85-NH Super Sabre (56-3400) 34th FDS, 413th FDW, George AFB, California 1959. To MASDC, Davis-Monthan AFB, November 1978, as #FE421. Converted to the QF-100D drone program #173.

F-100D-50-NH Super Sabre (55-2868) 454th FBS, 323rd FBW, Bunker Hill AFB, (Now Grissom AFB) Indiana 1957. Crashed February 1966, 6 miles SSW of Aviano AB (Italy) following engine failure.

F-100D-30-NA Super Sabre (55-3783) USAFE 390th TFS, 366th TFW, England AFB, Alexandria, Louisiana 1959. 366th TFW was deactivated in early 1959. Then with 613rd TFS, 401st TFW, in Vietnam.

F-100D-90-NA Super Sabre (56-3263) PACAF, 510th TFS, 405th TFW, Clark AB Luzon, Philippines 1962.

USAFE 49th TFW, Wing Commander's personal aircraft. Etain-Rouvres AB France 1958. Crashed 11 miles south of Torrejon AB, Spain 1962. Pilot ejected safely.

F-100D-30-NA Super Sabre (55-3722) 31st TFW, 9th Air Force, Weapons Team marking, George AFB, California 1957. Shot down by AAA 25 miles SW of Hue, Thua Thien Province, South Vietnam Jul 1969. Pilot ejected and was rescued by USMC helicopter.

F-100D-45-NH Super Sabre (55-2839) USAFE 493rd TFS, 48th TFW, Chaumont AB, France 1959. Crashed on

F-100C-1-NA Super Sabre (703) Ex 53-1709. To get more information on this F-100, read the JF-100C piece in this book.

North American TF-100C-1-NA Super Sabre, US Air Force, 54-1966, Prototype, Edwards Air Force Base California, ca. 1955

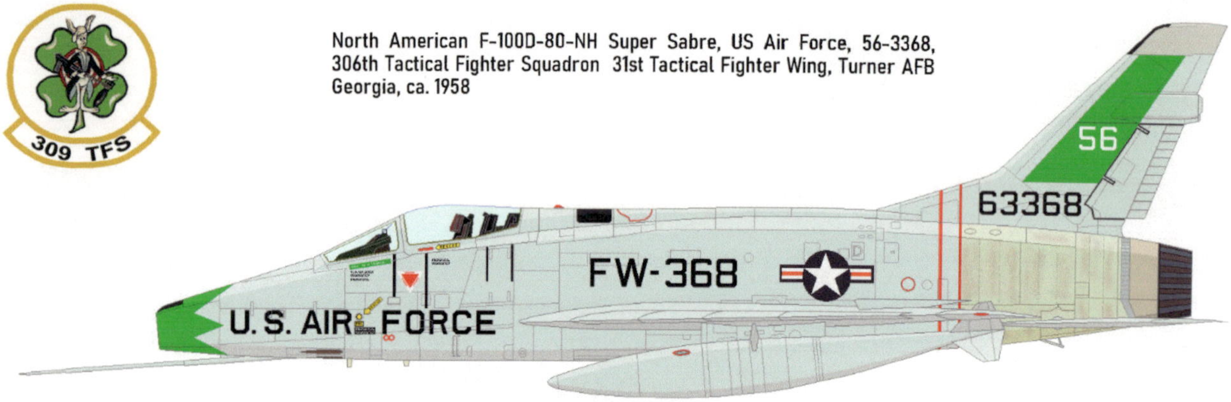

North American F-100D-80-NH Super Sabre, US Air Force, 56-3368, 306th Tactical Fighter Squadron 31st Tactical Fighter Wing, Turner AFB Georgia, ca. 1958

North American F-100D Super Sabre, US Air Force, 55-3668, 55th Tactical Fighter Squadron 20th Tactical Fighter Wing, RAF Weaterssfield United Kingdom, ca. 1959

North American F-100C-20-NA Super Sabre, US Air Force, 54-1999, 22nd Tactical Fighter Squadron 36th Tactical Fighter Wing, Bitburg AB Germany, ca. 1958

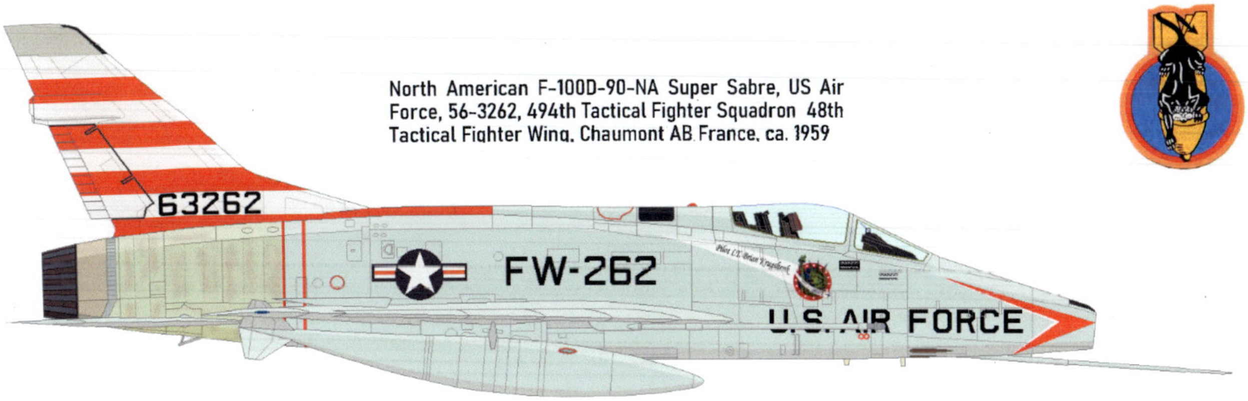

North American F-100D-90-NA Super Sabre, US Air Force, 56-3262, 494th Tactical Fighter Squadron 48th Tactical Fighter Wing, Chaumont AB France, ca. 1959

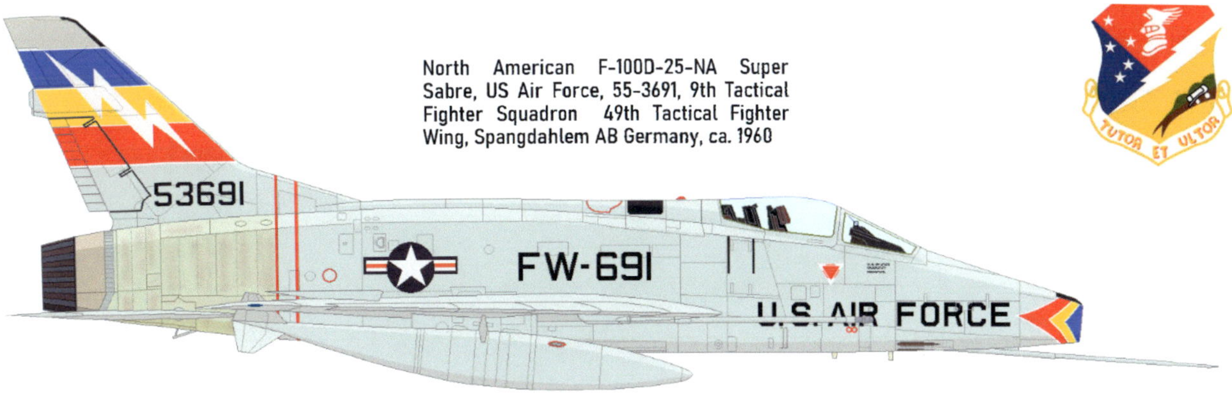

North American F-100D-25-NA Super Sabre, US Air Force, 55-3691, 9th Tactical Fighter Squadron 49th Tactical Fighter Wing, Spangdahlem AB Germany, ca. 1960

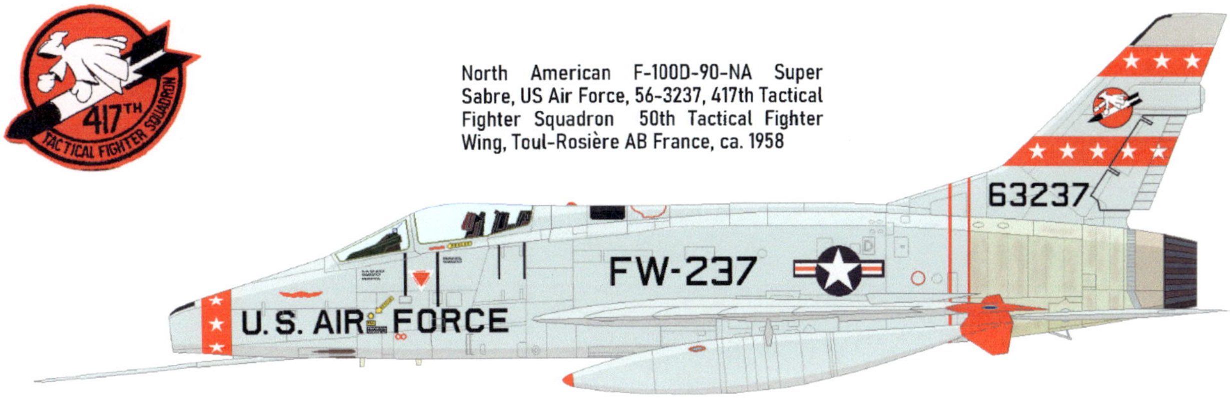

North American F-100D-90-NA Super Sabre, US Air Force, 56-3237, 417th Tactical Fighter Squadron 50th Tactical Fighter Wing, Toul-Rosière AB France, ca. 1958

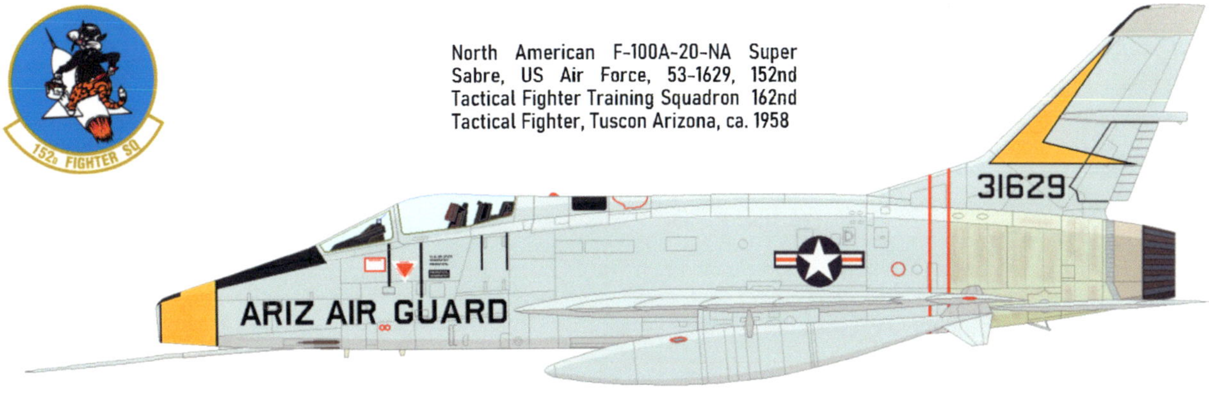

North American F-100A-20-NA Super Sabre, US Air Force, 53-1629, 152nd Tactical Fighter Training Squadron 162nd Tactical Fighter, Tuscon Arizona, ca. 1958

North American F-100C25-NA Super Sabre, US Air Force, 54-1888, 32 Tactical Fighter Squadron 36 Tactical Fighter Wing, Soesterberg AB Netherlands., ca. 1960

North American F-100D-65-NA Super Sabre, US Air Force, 56-2980, 492 Tactical Fighter Squadron 48 Tactical Fighter Wing , RAF Lakenheath United Kingdom, ca. 1961

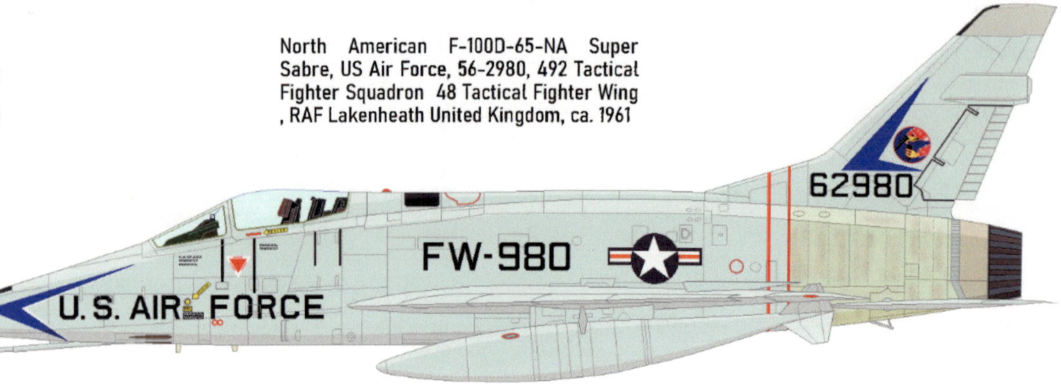

North American F-100D-30NA Super Sabre, US Air Force, 55-2913, 458 Fighter Bomber Squadron 506 Fighter Bomber Wwing, Tinker AFB Oklahoma. 1957

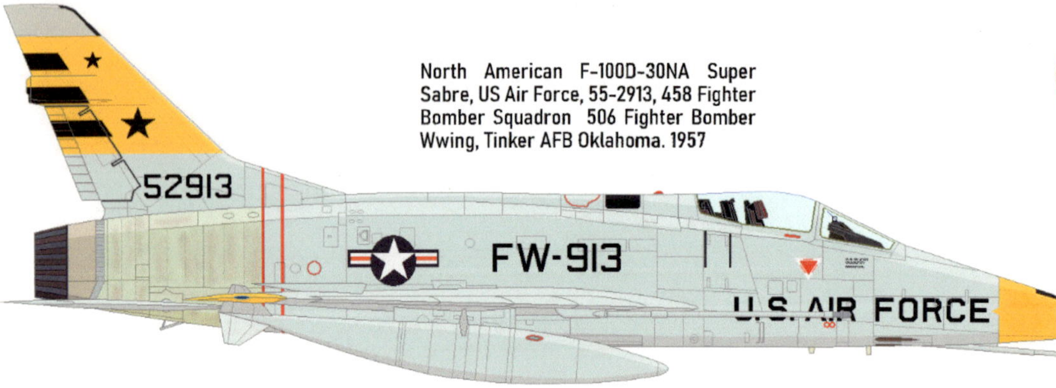

North American F-100D-30NA Super Sabre, US Air Force, 55-3375, 615 Tactical Fighter Squadron 401 Tactical Fighter Wing, England AFB Louisana. 1958

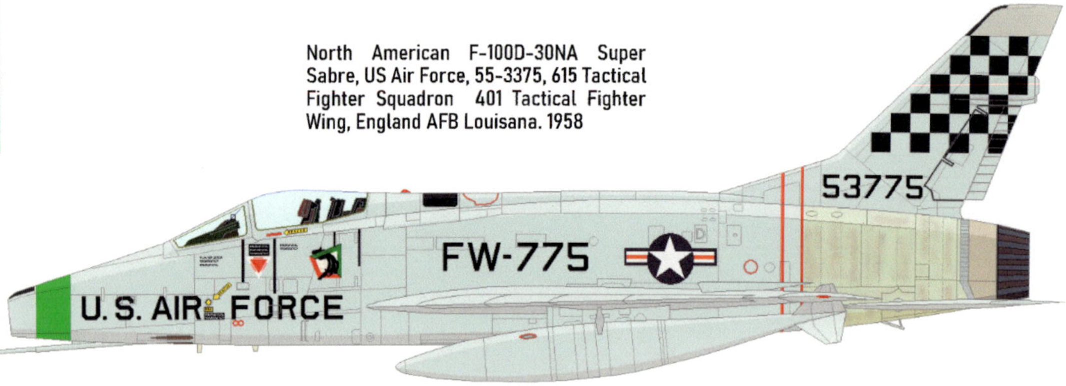

North American F-100F-10-NA Super Sabre, US Air Force, 56-3814, 417th Tactical Fighter Squadron 50th Tactical Fighter Wing, Hahn AB Germany, ca. 1965

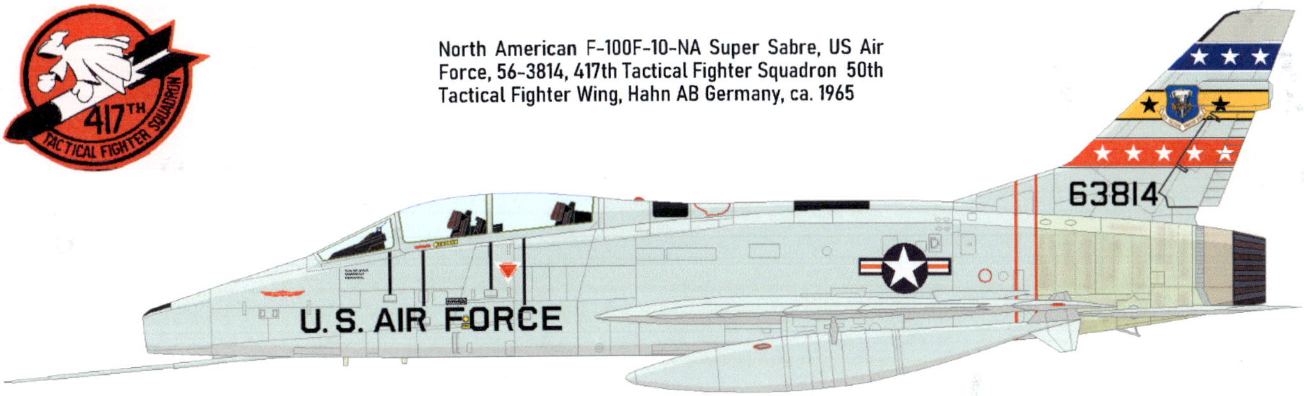

North American F-100D-90-NA Super Sabre, US Air Force, 56-3264, 510 Tactical Fighter Squadron 405 Tactical Fighter Wing, RTAFB Don Muang Thailand, ca. 1961

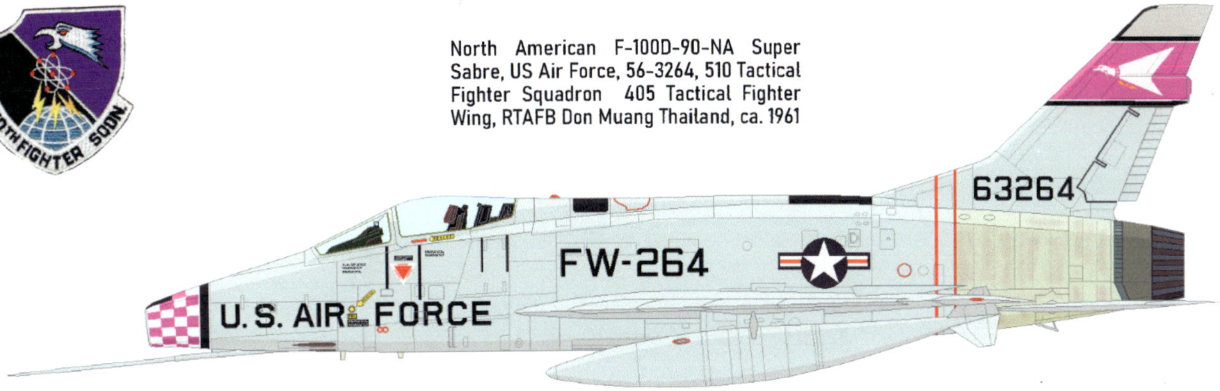

North American F-100C Super Sabre, US Air Force, 55-2720 7235 Service Squadron 7272 Flying Training Wing, Wheelus AFB Libya, ca. 1965

North American F-100D-25-NA Super Sabre, US Air Force, 55-3642, 522 Tactical Fighter Squadron 23 Air Base Group, Da Nang AB Vietnam, ca. 1964

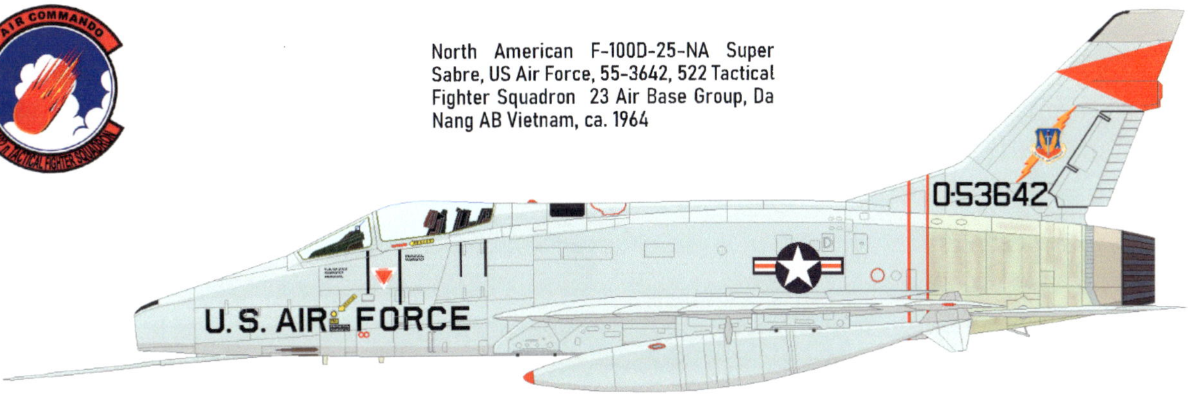

North American F-100A-1-NA Super Sabre, US Air Force, 52-5757, Prototype, Edwards Air Force Base California, ca. 1954

North American F-100F-15-NA Super Sabre, US Air Force, 56-3950, 1 Fighter Day Squadron 413 Fighter Day Wing, George AFB California, ca. 1958

North American F-100C-25-NA Super Sabre, US Air Force, 54-2101, 434 Tactical Fighter Squadron 479 Tactical Fighter Wing, George AFB California., ca. 1958

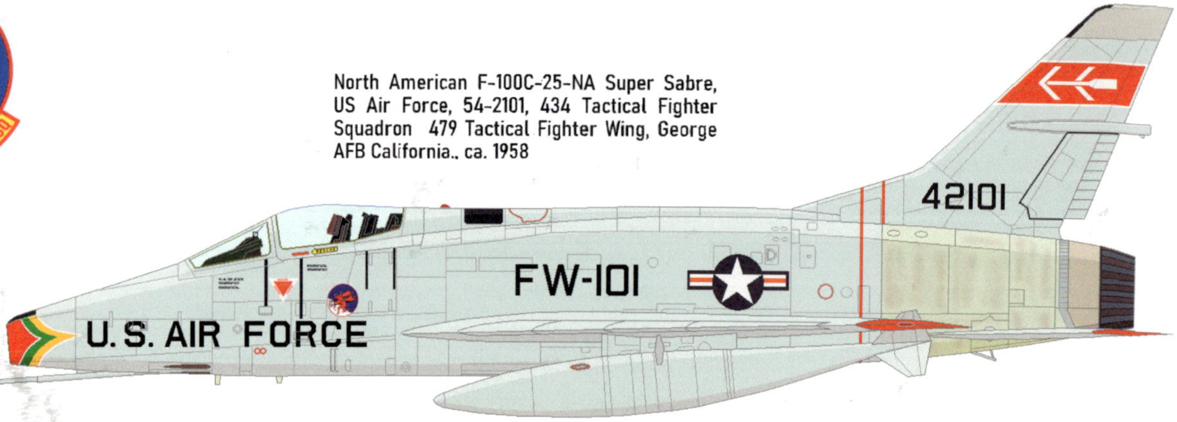

North American F-100A-10-NA Super Sabre, US Air Force, 53-1555, Air Training Command, Nellis AFB Nevada., ca. 1954

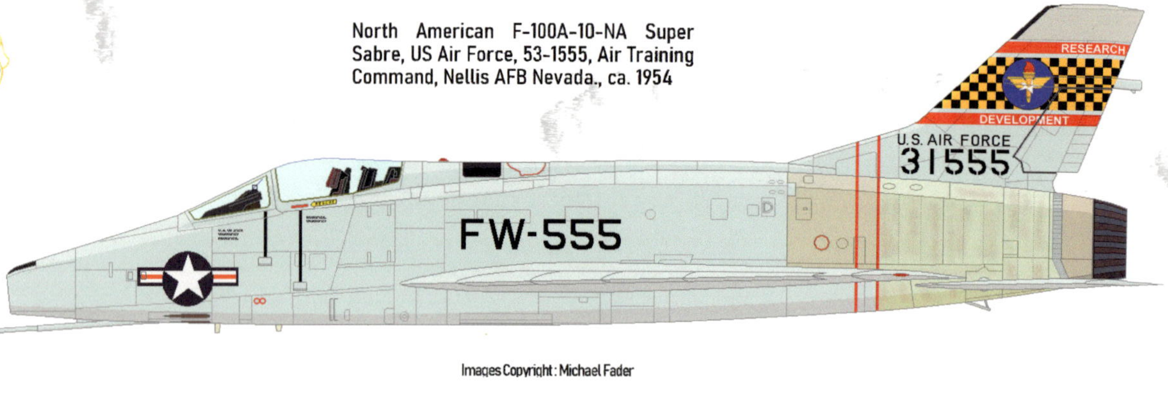

F-100D-60-NA, 56-29-28, in the 81st Tactical Fighter Sq., Tactical Fighter Wing, at Wheelus Air Base, Libya, for gunnery training in 1958. In May 1994, sent to Robins AFB, Georgia, and placed on a pedestal as a gate guard for the Base. (USAF)

Previously, weapons systems changed so often that each aircraft differed from the next. The High Wire modifications, which required about 60 workdays per aircraft, were implemented on about 700 F-100s. They concluded in mid-1965. The modifications cost $160 million. The project was successful and consolidated the 'C' and 'D' versions.

Even after the High Wire project's success, the F-100D still experienced more problems. Several accidents resulted from landing-gear malfunctions and unreliable drag chutes. Often, the parachute failed to deploy correctly or, worse, broke free, causing the plane to overshoot the runway. Furthermore, while compressor stalls in the J-67-21 engine continued, the new source of worry was the engine itself, which had gone past the point at which it could be reliably serviced, and for which there was no viable replacement. This extra problem was solved after a complete engine overhaul, although it was not completed without difficulties. Some of the rebuilt J-67 engines were utterly inadequate. Moreover, a chronic shortage of spare parts further delayed the F-100 Inspect, Repair As Needed (IRAN) cycle. The F-100s' troubles kept the mechanics busy.

In 1966, the Air Force decided to prolong the service life of its F-100s. To do this, the USAF investigated the extent and costs of structural changes required to lengthen the potential serviceable life of the aircraft to 5,500 hours, against the 3,000 hours suggested by the manufacturer, and later changed again to 7,000 hours to permit the retention of the 'D' and 'F' model series past 1971. Because of the F-100's high rate of weapon deployments in Southeast Asia, the USAF, with assistance from NAA, examined the structural state of the aircraft's wings in 1967. They also considered modifying the skin beneath the wings if necessary, given the aircraft's extensive combat service, which had caused wear and tear in that area. The urgent need for safety upgrades was confirmed in mid-1967 when an F-100 crashed due to wing failure. After the incident, the Air Force grounded several F-100s until it reinforced their wings. The Air Force then developed a comprehensive F-100 structural upgrade program in late 1967. After a lengthy modification process, the Air Force completed the middle wing sections on 682 F-100D and 'F' versions by 1969. The process continued, with 200 more aircraft remaining. The Air Force also completed the alteration of the F-100's lowest outer wing panel, which it deemed critical before the aircraft reached 4,000 flight hours.

During the F-100D's service lifetime, the USAF lost around 500 aircraft, primarily to accidents, accounting for roughly 40% of the overall F-100D fleet. Most of the mishaps occurred within the aircraft's first two years of active service. After that, the Air Force frequently modified the F-100Ds, and pilots underwent a comprehensive training program to reduce pilot error and technical problems. These two moves significantly reduced accident rates. Unfortunately, this pattern reversed as F-100D combat missions intensified in the late 1960s during the Vietnam War. As expected in a combat scenario, the number of F-100s lost rose.

In 1962, the 20th Fighter-Bomber Wing, based at Shaw Air Force Base, S.C., was reequipped with North American F-100 Super Sabres, such as the F-100D-25-NA, 55-3644, which were suited for nuclear 'toss bombing' missions because of their greater power and speed. (USAF)

F-100F

The next F-100 version after the 'D' version was the F-100F-1-NA (56-3725), which first flew on March 7, 1957, piloted by Alvin S. White. The aircraft took off from Los Angeles, where the Air Force manufactured all 'F' versions. The converted TF-100C served as the F-100F prototype.

Although the F-100F was fully equipped for operational duty, it was limited in terms of armament. Instead of the 4, Pontiac built M39 20 mm cannons that the single-seat F-100s had. The 'F' variant had only two because of its extended cockpit. Moreover, its weapons capacity was only 2,700 kg. It was mainly used as a trainer and fighter-bomber. Like all F-100s in service, it carried extra fuel tanks for Buddy refueling.

Because the F-100F was substantially similar to the F-100D, it could speed up the aircraft's operational testing phase. As a result, the tests were quite limited in performance and quality. They were completed in May 1958. However, the production run for the 'F' model was short. As with the F-100D, the Air Force slowed production of the F-100F. Then, in October 1959, the Air Force halted all F-100F production. The total production of the type was 339 units.

Although the F-100F was fully equipped for operational duty, a key difference hampered it. Instead of the 4 20 mm cannons the single-seat F-100s had, Pontiac built M39 20 mm cannons; the 'F' variant had only two. Despite this, it could carry a 6,000 lb. ordnance load. During the Vietnam War, the F-100F saw significant action. It participated in many combat operations, mainly as a fast-moving Forward Air Control and Electronic Warfare aircraft. Like all F-100s in service, it carried extra fuel tanks for Buddy refueling. The F-100F was the last production model of the F-100 family. It brought total F-100 production to 2,294. The last F-100 delivered was in

This is an F-100F-10-NA, s/n 56-3890, of the 48th Tactical Fighter Squadron, at RAF Lakenheath, UK. The Turkish AF purchased it on September 8, 1972. (USAF)

This is a privately owned F-100F. (wallycacsabre)

This is an F-100F-20-NA, s/n 58-1232, seen here at Phan Rang Air Base in South Vietnam. This is a Wild Weasel aircraft. Assigned to the 510th TFS, attached to the 405th FW. Later assigned to the 27th TFW, then to the 352nd TFS, 35th TFW during the Vietnam War. It was then allocated to the 120th TFS, Colorado ANG. Then moved to the 175th TFS, South Dakota ANG. It spent some time on display at Grissom Air Museum at Grissom AFB, Indiana. In Oct. 1987, it was allotted to a pedestal at Brooks AFB, TX, marked 56-3944.
It then moved to the Warner Robins Air Museum. It was sent to the Pacific Aviation Museum, Ford Island, Hawaii. It ended up at the Military Aircraft Storage and Disposition Center (MASDC) on Dec. 9, 1971, as FE123.

ZELL

The F-100s were used as lab rats in the 'ZELL' experiment (Zero Length Launch) in the late 1950s. The ZELL concept was based on the idea that airplanes could launch from within a heavily protected aircraft shelter using a reusable rocket booster. The theory emerged in reaction to the common belief that planes would need to be propelled into flight if a nuclear attack made airfields and runways unusable. The ZELL mission was designed to be a one-way mission. Once in the air, a pilot was supposed to fly to the target (over or near enemy territory) and deliver his nuclear bomb. Then he had to find a safe place to land or eject, which was not easy over hostile territory. Basically, the pilot's odds of success were minimal. The Air Force first used the Republic F-84 as a test aircraft in the early 1950s. Then, in 1958, the F-100s were tasked with testing ZELL.

On March 26, 1958, NAA test pilot Albert W. Blackburn took command of an F-100D deployed on a truck-mounted launcher. He was the first pilot to test the apparatus with the F-100. Afterward, more tests were performed. The F-100 Super Sabres were fully equipped when they were launched straight into the sky by a vast solid rocket, either a Rocketdyne M34 with 130,000 lb. thrust or an Astrodyne unit with 150,000 lb. thrust. The rockets gave the F-100 an immediate acceleration of 4 G. The aircraft reached 320 mph after four seconds of sustained diagonal bursts. Unfortunately, these experiments were not very successful. The heavy engines that had to be attached to the aircraft, the need to lift the plane on the truck launcher, and the pilot having to enter the cockpit via a very long ladder made logistics a nightmare. Everything was just too awkward. Because of logistical issues, the Navy eventually canceled all ZELL tests. Thankfully, the increasing effectiveness of guided nuclear missiles, such as the Polaris A-1, which entered service in December 1959, negated the need to continue the ZELL experiments, making such aircraft deployment less essential for tactical strategists.

A fully equipped F-100 being propelled to the sky at 4 G by the enormous ZELL rocket engine. The uncanny ZELL project was awkward to set up and execute. (USAF)

F-107 ULTRA SABRE

The F-107A flies over rural America. It had a unique, interesting shape. (USAF Museum)

In October 1953, NAA completed four months of engineering study and launched its new project, the NA-211, which was the company's designation for the F-100B. The aircraft was to be an all-weather interceptor. In 1954, North American offered the Air Force, specifically the Strategic Air Command (SAC), the F-100B, a more powerful version of the F-100A. The Air Force was not impressed with the design. Furthermore, it told NAA it was interested in Convair's new interceptor, not NAA's F-100B, and that Convair would manufacture the Air Force's interceptors. It was as if NAA had been struck by lightning. All the work it had put into designing the new interceptor was for nothing. However, the USAF did offer NAA some hope. It told NAA that the Republic F-105 could use some competition, and it ordered three NA-212 prototypes (55-118 to 55-120). The USAF told NAA that, to succeed, it would have to implement significant alterations to the F-100B model.

NAA began implementing the various modifications to win over the USAF. However, the changes were so numerous that the plane received a new designation, YF-107A. As a result, NAA and the USAF signed a contract in August 1954 for three aircraft. Later, the Air Force ordered six additional F-107As but canceled the order. NAA had offered the USAF two versions of the F-107A. The first was the original all-weather interceptor version, NA-211, which had no chance of success because the Air Force preferred the Convair interceptor over the F-107A. The second was the ground-attack variant designated the NA-212, and the third was named 'Sea Fury.' This aircraft was not aimed at the USAF but the US Navy. However, as it never got a company number, it was never heard of again. So, the N212 was NAA's only real option for the Air Force.

NAA needed to correct fundamental errors in the F-100's initial design. The intake was the first area to draw significant visual attention. Because of the need for radar, the inlet had to be relocated from its original nose location to a new position above the fuselage. At that position, pressure recovery was outstanding, even at steep angles of attack.

The wing was derived from the F-100, but with integrated tanks and after being entirely re-tensioned to support heavier weights. The trailing edge was beefed up with trim tabs and slotted flaps. However, the leading edge saw little alteration. Spoilers above and below the wings provided lateral control. The second intriguing element was the use of a single-piece slab vertical surface.

The F-107's powerplant was the Pratt & Whitney J-75-P-9, which was introduced in 1955. It was an axial-flow, dual-compressor turbojet engine producing 16,100 lb. thrust and 24,500 lb. thrust with afterburner. The F-107 also had provisions for 10,000 lb. of underwing stores and four 20 mm cannons. The USAF built a total of three F-107s. The first example, 55-518, flew on September 10, 1956. The other two followed shortly after: 55-5119 flew on November 28, and 55-5120 on December 10.

A 'fly-off' between the F-105 Thunderchief and the F-107 Ultra Sabre' was planned by the Air Force in September 1956 to gauge the performance of both aircraft. Based on the specifications, the F-107 was smaller than the Republic F-105 Thunderchief. It measured 61 feet 10 inches long, had a 36-foot 7-inch wingspan, and had a maximum takeoff weight of 41,537 lb. The F-107 had a top speed of 1,295 mph, a range of 2,428 miles, and a service ceiling of 53,200 feet. The F-105 Thunderchief was 63 feet 1 inch long, had a 34-foot 11-inch wingspan, and a maximum takeoff weight of 46,998 lb. It could reach 1,367 mph and had a maximum range of 2,228 miles.

Although the F-105 had a slight edge in speed and takeoff weight, the F-107 was a solid competitor for the Thunderchief. North American was confident it would win the USAF contract. Unexpectedly, the Air Force told NAA it had to cancel the fly-off because the F-105 was not ready. The 'Thud' had many teething problems early in development, including two crashes. After canceling the 'fly-off,' the USAF stated it would purchase the F-105. The F-107's fate had already been sealed six months earlier. In March 1956, the USAF ordered 65 F-105Bs and 17 RF-105Bs.

Many aviation pundits were surprised by the Air Force's decision. They believed the F-107 had the edge over the F-105 and should have taken the honors. Many thought the F-107's higher rate of climb and service ceiling surpassed the F-105's. Moreover, the F-107's numerous advanced features, including enhanced longitudinal control surfaces, a one-piece movable vertical stabilizer, and roll control via spoilers rather than ailerons, were valid reasons the USAF should have chosen it.

Despite its exciting design and excellent performance, the F-107 was a complete failure. The cynics said the Air Force chose Republic's aircraft not because the 'Thud' was a better plane, but because Republic needed work; its production had dropped significantly in the mid-1950s, with no new projects in the pipeline except the F-105. The skeptics may have been right, but the 'Thud' proved to be a suitable aircraft in SEA. It could carry 4,000 pounds more ordnance than the F-107 and used the Pratt & Whitney J-75 engine. The F-107 Ultra Sabre' was a well-designed and well-equipped aircraft, but it wasn't meant to be. Its permanent home now is at the USAF Museum at Wright-Patterson Air Force Base near Dayton, Ohio.

A left-side profile of the F-107A at the USAF Museum. (National Museum of the USAF)

An outstanding photo of the F-107A. (National Museum of the USAF)

The F-107A is being towed through the museum grounds. (National Museum of the USAF)

The cockpit of the F-107A. ((National Museum of the USAF)

The F-107A "Ultra Sabre" in flight (National Museum of the USAF)

F-100 OPERATIONAL HISTORY

In June 1955, the first USAF unit to receive the F-100A Super Sabres was the 479th Group (434th, 435th, and 436th Squadrons). It was followed by the 450th Group (720th, 721st, 722nd, 723rd Squadrons) and the 322nd Day Fighter Group (450th, 451st, 452nd Squadrons).

The F-100's operational use peaked in 1957, when the USAF had 16 Super Sabre units. From then on, the USAF gradually phased out the F-100A and F-100C and transferred them to Air National Guard units. However, in 1964, the USAF still had 10 active F-100 wings, all equipped with F-100Ds.

The 10 Tactical Air Force Wings are in red, and their squadrons are in blue. The TFWs based in the United States, were as follows: 3 TFW (90, 416, 510, 531) England AFB, 27 TFW (478, 522, 523, 524) Cannon AFB, 31 TFW (306, 307, 308, 309) Homestead AFB, 354 TFW (352, 353, 355, 356) Myrtle Beach AFB, 401 TFW (612, 613, 614, 615) England AFB, 474 TFW (428, 429, 430, 481) Cannon AFB.

The following were outside the United States: **20 TFW** (55, 77, and 79) in the UK; 8 TFW (492, 493, 494) in the UK; 50 TFW (10, 81, 417) in Germany; and **405 TFW** (511) in the Philippines.

Meanwhile, the ANG had 11 squadrons equipped with the older F-100A & C versions. They were as follows: 152 FIS Tucson, Arizona ANG, 188 TFS Albuquerque, New Mexico ANG, 110 TFS St. Louis, Missouri ANG, 118 FIS Hartford, Connecticut ANG, 119 TFS Atlantic City, New Jersey ANG, 120 TFS Denver, Colorado ANG, 121 TFS District of Colombia, Washington DC ANG, 127 TFS Wichita, Kansas ANG, 136 TFS Niagara Falls, New York ANG, 166 TFS Lockbourne, Ohio ANG, and 174 TFS Sioux City, Iowa ANG. After almost five years on the front lines, the F-100s were gradually replaced by the F-105 Thunderchiefs and F-4 Phantoms. The F-100 Super Sabres were distributed to 19 ANG Squadrons, which used them until 1980.

F-100F-15-NA, 56-3992, of the 20th TFW, at RAF Wethersfield, UK, June 11, 1966. This aircraft was a project 'Hi-Wire' conversion. Transferred to the Turkish AF on Nov 11, 1977, as 56-3992. This aircraft was W/o on July 19, 1983, due to a fatal crash near Konya AFB, Turkey (USAF)

F-100D Super Sabre 56-3410 of the 494 TFS, 48 TFW, in 1965. Later sent to MASDC on January 18, 1977, as FE302. Converted to QF-100D drone No. 135. W/O 22 January 1985. Shot down by an AIM-9P missile. (RuthAS)

Three 353rd Tactical Fighter Squadron aircraft line up to get refueled by a Boeing KB-50 aerial tanker. (USAF)

F-100D-85-NH 353rd TFS in 1960. to MASDC May 9, 1979 as FE522. Converted to QF-100D drone no. 367. W/O 11 Feb.1992. Shot down by AIM-7M missile. (USAF)

353rd Tactical Fighter Squadron F-100 Flightline 1960. (USAF)

F-100 Super Sabre, 56-3386, of the 353rd TFS over snowy mountains. (USAF)

353rd TFS F-100, 56-3388, Wingman in 1960. Later, with the 615 TFS. This F-100 struck the approach lights after landing short of the runway at night on Oct. 25, 1970. (USAF)

F-100D-25-NA, 55-3679, of the 79th Tactical Fighter Squadron, 20th Tactical Fighter Wing, RAF Woodbridge, Suffolk (UK), at RAF Waterbeach, Cambridgeshire, on Battle of Britain Day, 14 Sep. 1963. Retired to MASDC on February 6, 1977, as FE0313. (TSRL)

F-100D-85-NH Super Saber, 56-3386, of the 353rd TFS over the Alps. (Wikimedia -source unknown)

Front view of an F-100D. Note the large 'elevator door' sized air brakes. (USAF)

F-100D, 56-3040, of 429th TFS, 3rd TFW, on the way to a target in Vietnam in December 1965. Later, with the 510th TFS, 3rd TFW. Shot down by small arms fire near Loc Ninh 60 miles North of Saigon, South Vietnam. Nov. 19, 1967. Pilot was killed. (USAF)

This QF-100D Super Sabre Drone was sent to MASDC on April 17, 1979 as FE501. Converted to YQF-100D drone no. 092, later to a QF-100D. W/O 29 May 1985. Shot down by an AIM-7M missile. (National Archives)

F-100C-5-NA, 54-1807, 121st Tactical Fighter Squadron. Aircraft retired to MASDC on Dec. 6, 1971. Transferred to the Turkish AF in June 1973. W/O May 12, 1974. (USAF)

F-100D-90-NA, 56-3328, of the 405th TFS, landing at Clark Air Base, Philippines. To MASDC on March 14, 1979, as FE476. Converted to a QF-100D drone No. 131. W/O June 4, 1985. Shot down by an AIM-7M missile. (USAF)

F-100D-80-NH, 56-3357, of the 352nd TFS, 354th TFW, Myrtle Beach AFB. (USAF)

F-100D-80-NH of 353rd Tactical Fighter Squadron 56-3352 in 1963. (USAF)

F-100D-30-NA, 55-3797, of the 416th TFS, at Bien Hoa Air Base, South Vietnam, in July 1965. The aircraft was retired to MASDC on February 2, 1979, as FE0456. Later, it was converted to a QF-100D drone, no. 152. It was W/o on June 11, 1985, when it was shot down by an AIM-9P missile. (USAF)

Pacific Air Forces (PACAF) F-100D-30-NA, 55-3765, from the 531st TFS, 21st TFW, on air defense patrol over Japan, in 1959. This aircraft was shot down by ground fire near Ban Ti Srenh, South Vietnam, on Jan. 4, 1968. The pilot ejected and was rescued by helicopter. (USAF)

118th Fighter Interceptor Squadron, F-100A-15-NA, 53-1601. Sent to MASDC on February 9, 1966. Later sold to Republic of China AF as 31601, code 0222. Eventually, it was WFU. (USAF)

F-4 Phantoms knocked out the Thanh Hoa Bridge with laser-guided weapons in 1972. (USAF)

The F-100 'Hun' in SEA

A bridge too far....

Some of the early targets in the Vietnam War for the Air Force were two bridges. Not just any bridges, but pain-in-the-neck bridges.

The bridges in question were hampering American efforts to stem the flow of fighters coming into South Vietnam from the North. The first was the Paul Doumer Bridge, on the outskirts of Hanoi. However, the second bridge was a major thorn in the Air Force's side. That was the Thanh Hoa Bridge. It was practically indestructible. The two bridges enabled the passage of men and materiel from North Vietnam to battle areas in Laos and South Vietnam. The constant flow of men and equipment through the bridges to infiltrate the South was significant, and destroying them would have been a major blow to North Vietnamese forces.

General William C. Westmoreland, commander of US forces in the Vietnam War from 1964 to 1968, saw an opportunity. He was a great exponent of interdiction raids, and the USAF was the main player in that field. The Air Force could take out the bridges quite efficiently, with few lives lost, or so it seemed. General Westmoreland felt that would be the case. He wanted to ensure that the interdiction would send a strong message to the North Vietnamese leadership that it was not going to be an easy ride into South Vietnam anymore. The General was convinced the war would be fought on the ground, but the US and South Vietnamese forces could only succeed with substantial Air Force support.

The attacks on the bridges began on April 3, 1965, and from the start, the Air Force focused on the Thanh Hoa bridge. A total of seventy-nine US aircraft participated in the assault on the bridge with various weapons, but failed to destroy it. The strikes on both bridges were difficult because they were at the end of the interdiction zone and required much more effort than Westmoreland and other ground-force leaders had anticipated. However, that would not be easy. The area around the bridges was teeming with enemy soldiers hiding under tree cover. Furthermore, the area was saturated with defense systems such as Soviet built surface-to-air missiles, anti-aircraft artillery, and enemy aircraft. The Air Force was confident it could destroy the Thanh Hoa bridge quite rapidly. That was not the case; in the initial attack, the Soviets destroyed an F-100F 'Wild Weasel' flak suppressor and an RF-101 Recon 'Voodoo'. Even worse, three 'Thuds' were shot down during the second assault, and one other F-105 was lost during the third raid. Luckily, the fourth strike resulted in no casualties. All the planes involved in the bridge attack were based in Northeast Thailand and flew about 380 miles to the target zone. They tried their best to destroy the bridge, but it held firm. The Air Force could not apply enough force to blast it to smithereens. In 1966, the North Vietnamese strengthened the Thanh Hoa Bridge's defenses even further. They surrounded it with more SAM installations. Flying through that area was like running the gauntlet. The area was too dangerous for close-in attacks. Destroying the bridge would not come easy. The participation of the 'Wild Weasel' F-100F in the initial strike on the bridge was one of the first battles in Vietnam for the 'Huns'. As for the bridge that would not die, it was many years later, in 1972, before USAF

F-4 Phantoms destroyed it with new precision electro-optical and laser-guided bombs. It was a remarkable feat for the North Vietnamese to keep the bridge open for so many years under heavy firepower from the USAF.

The F-100's operational life, first designed as a fighter-bomber and later as a nuclear weapon delivery platform, was ending by the early 1960s. The new guided nuclear missiles were more efficient than the F-100's quaint 'Low Altitude Bombing System' (LABS), which flung a nuclear bomb at the target while the F-100 made a run for it before detonation. That system was already antiquated in 1960. It was just too awkward and inefficient. The F-100 was nearing the end of its tenure. It was no longer a pure fighter or a nuclear-weapon bomber. It was a fighter-bomber with four powerful M39 cannons, while contemporary fighters and interceptors carried various missiles. Its days were numbered; the Hun had no work. No one was left to fight, and no one left to bomb. Then, the Vietnam conflict broke out and saved the day.

The F-100 was constantly being upgraded. The last 444 F-100Ds, built at the Los Angeles Inglewood factory, carried AIM-9B Sidewinder air-to-air missiles. Finally, the Hun had its own missiles. All previous F-100Ds, and some Cs, were retroactively modified as well. However, the biggest changes went to the F-100Ds and Fs, as they prepared to head to the sunny skies of Vietnam. The D and F variants received hundreds of modifications. The major mods focused on the aircraft's structure and equipment. They also added many hi-tech mods, including active and passive Electronic Countermeasure (ECM) equipment, a supersonic autopilot, a new IFF (Identification Friend or Foe), and UHF radios. There was also the camouflage. No F-100 pilot would have felt safe sitting in a shiny titanium fuselage, flying over SEA, reflecting bright sunlight and letting the VC know exactly where to fire the next SAM missile. The camo suited the F-100 very well.

In June 1964, at the Laos government's request, the US Navy conducted recon flights along the Laos-Vietnam border.

The Vietnamese did not like what they saw and shot down two US Navy RF-8 Crusaders, the reconnaissance version of the F-8 Crusader.

The unwarranted attack on the RF-8s by the North Vietnamese prompted an immediate retaliation against North Vietnamese anti-aircraft positions, which was successfully carried out by F-100s based at Clark AFB in the Philippines. In fact, the June 1964 attacks on the Vietnamese positions were not the first time the F-100s saw action in Southeast Asia (SEA). That happened in May 1962, when several 'Huns' were sent to Thailand from the 13th Air Force in response to communist incursions into northwest Laos. The F-100s went in and took care of the targets.

The 3rd, 31st, 35th, and 37th Tactical Fighter Wings operated F-100s from bases in South Vietnam in the late 1960s because the F-100 was an important component of close air support in Vietnam. This meant that in 1967 only five F-100 squadrons remained in the United States, which profoundly affected the Air National Guard.

This is a 162nd Tactical Fighter Squadron F-100D-90-NA, 56-3255. The aircraft was sent to MASDC on April 6, 1978, as FE0384. It was later converted to a QF-100D drone and expended. (USAF)

North American F-100D-75-NA, 56-3184, dropping a 500 lb. Snake-Eye bomb on a suspected Viet Cong position in 1966. This aircraft was retired to the MASDC on 23 August 1979 as FE0589. It was later converted into a QF-100D target drone. (USAF)

The ANG was due to be reequipped with the F-100s in the late 1960s; however, that plan was postponed because the F-100s in Vietnam had to remain there to keep the combat squadrons fully operational. They were vital to close air support. For example, close air support aircraft flew 800 sorties every day in Vietnam, but the F-100s were crucial to these missions, as they performed most of the bombing.

Whilst on these assignments, the Huns carried Mark 81 Snakeye bombs fitted with a Mark 14 TRD (Tail Retarding Device), which slowed the fall of the bomb. The Snakeye bombs were perfect for close air support roles, as they allowed for very low-level bombing runs by the Huns. Apart from the Snakeye bombs, the Huns also carried cluster bombs, napalm, and 2.75-inch rockets. Whilst going down low for a bombing run, the F-100s were in danger of getting hit by small arms fire, so the F-100 pilots were advised to break their attacking dives at an altitude of 2,000 feet to avoid a hail of bullets from the Vietcong. However, many pilots flew much lower, risking being hit. Whether that was bravado or overconfidence, it was dangerous.

Between 1964 and 1969, the four F-100 squadrons stationed in South Vietnam flew three of four support missions, totaling over 300,000 sorties. During those years, each F-100 flew at least one mission per day, and the availability rate stayed above 80%, with refueling and preparation for a new mission often taking less than 30 minutes.

The operations were mainly close support attacks with the use of conventional bombs, such as napalm, the use of the M.39 guns, and rocket strikes. As for rockets, the highly touted Bullpup missiles did not live up to expectations. For instance, in the infamous Thanh Hoa Bridge assault on April 3, 1965, 16 F-105s lined up to fire their Bullpups at the bridge. Because the Bullpups had to be manually guided to the target, each Thunderchief had to line up for attack twice, in separate passes. After the F-105s fired their Bullpup missiles, the pilots noticed the bridge was essentially undamaged. The Bullpups were bouncing off the bridge like basketballs, without exploding. Consequently, their lack of destructive power, the need to make separate passes for each release, and the need to keep guiding the weapon through its flight led the USAF to conclude the Bullpups were inadequate. That was a euphemism for them being absolute duds. The USAF phased out the Bullpup from service in the 1970s.

122nd Tactical Fighter Squadron F-100D-75-NA, 56-3171. Retired to MASDC on March 9, 1979, as FE0525. Later, it was converted to a QF-100D target drone. (USAF)

The F-100s that participated in the Vietnam War were predominantly the F-100D and 'F' variants. There were also a few 'C's involved. The F-100s took on many roles in Southeast Asia, including low-level ground attack, electronic-warfare platforms (Wild Weasel), and 'fast-mover' FAC (Forward Air Control) aircraft. Day or night, the four Tactical Fighter Wings operating in Vietnam took off in their sturdy F-100s, mainly on close-support missions. The F-100 remained the most widely used close-air-support aircraft during the Vietnam conflict. The USAF generally phased it out of service in 1972. Throughout the F-100's service life, 889 Huns were destroyed in accidents, killing 324 pilots. The worst year for F-100 accidents was 1958, when 116 F-100s crashed, and 47 pilots were lost. However, the big record for the 'Hun' was that after nearly eight years of air combat in Vietnam, it flew a total of 360,283 combat sorties, more than any other USAF attack aircraft. Moreover, the North American Aviation F-100s flew more sorties in Vietnam than the 16,000 North American P-51 Mustangs did throughout the Second World War.

Maybe the first Century fighter wasn't a 'genuine' fighter jet, but as a close air support aircraft, it had no competition. In Vietnam, it did its duties faithfully and efficiently. The Hun might not have been as glamorous as other 'Century' jets, such as the Convair Delta jets and the F-104 Starfighter, but it was the first jet to break the sound barrier in level flight and paved the way for the rest of the 'Century' jets.

An F-100D Super Saber (s/n 56-3101) from the 429th Tactical Fighter Squadron, 474th Tactical Fighter Wing, on temporary duty at Tan Son Nhut Air Base, South Vietnam, in 1965. This aircraft was retired to the MASDC on July 6, 1978, and designated as FE0392. (USAF, Wikimedia)

An F-100D-85-NH Super Saber nicknamed "Snooper" of the 90th TFS, 3rd TFW, based at Bien Hoa Air Base, South Vietnam, circa 1968. (Aaron Headly-Flickr)

North American F-100D-25-NA Super Sabre, US Air Force, 55-3678, 493rd Tactical Fighter Squadron 48th Tactical Fighter Wing, RAF Lakenheath United Kingdom, ca. 1971

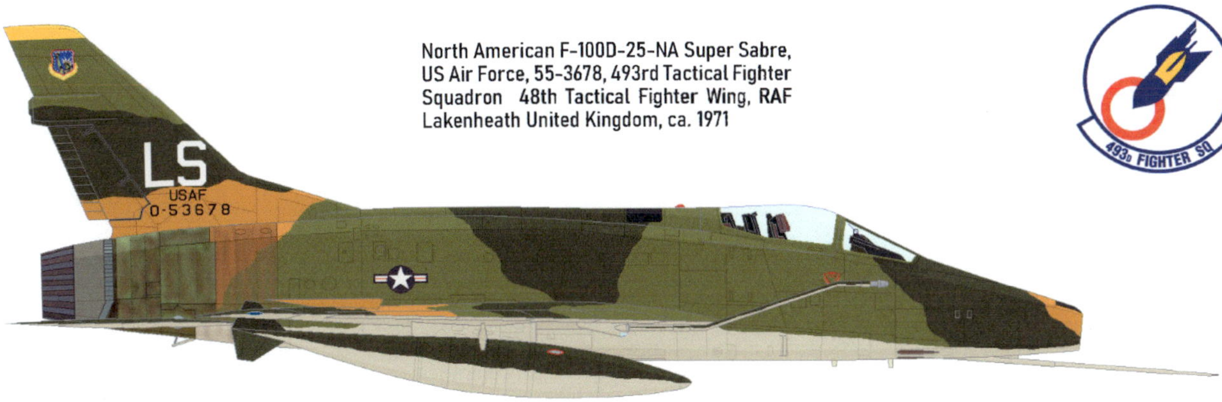

North American F-100D-35NA Super Sabre, US Air Force, 55-3604, 90th Tactical Fighter Squadron 3rd Tactical Fighter Wing, Bien Hoa Air Base Vietnam, ca. 1966

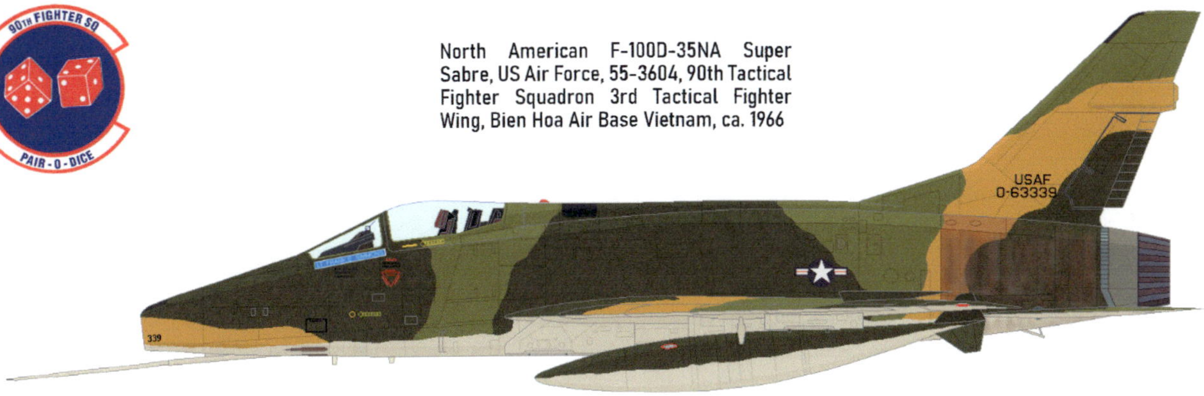

North American F-100D-60-NA Super Sabre, US Air Force, 56-2910, 531st Tactical Fighter Squadron 3rd Tactical Fighter Wing, Bien Hoa Air Base Vietnam, ca. 1969

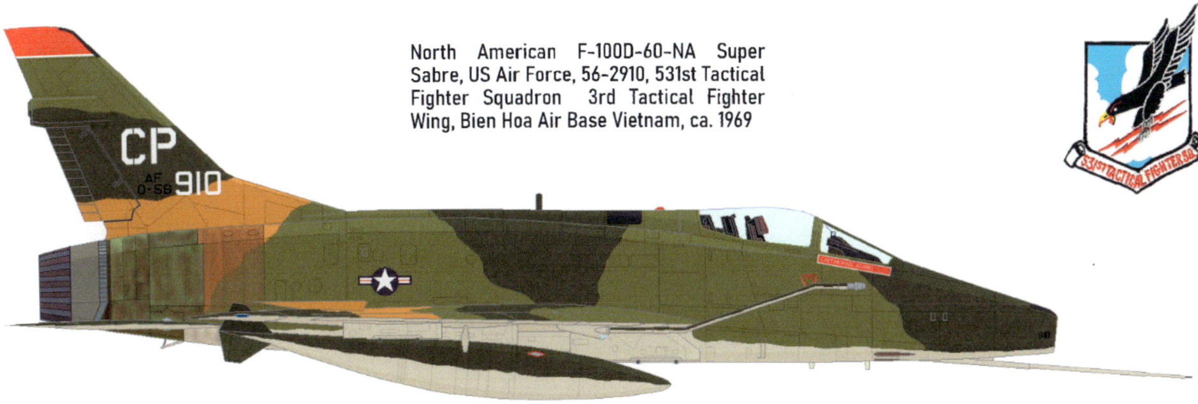

North American F-100F Super Sabre, US Air Force, 56-3760, 107th Tactial Fighter Squadron 127th Tactical Fighter Groop, Selfridge Air Force Base Michigan ca. 1973

North American F-100F-20-NA Super Sabre, U.S. Air Force, 58-1226, Det. Weasel 6234th Tactical Fighter Wing (Provisional), Royal Thai Air Force Base Khorat Thailand, ca. 1965

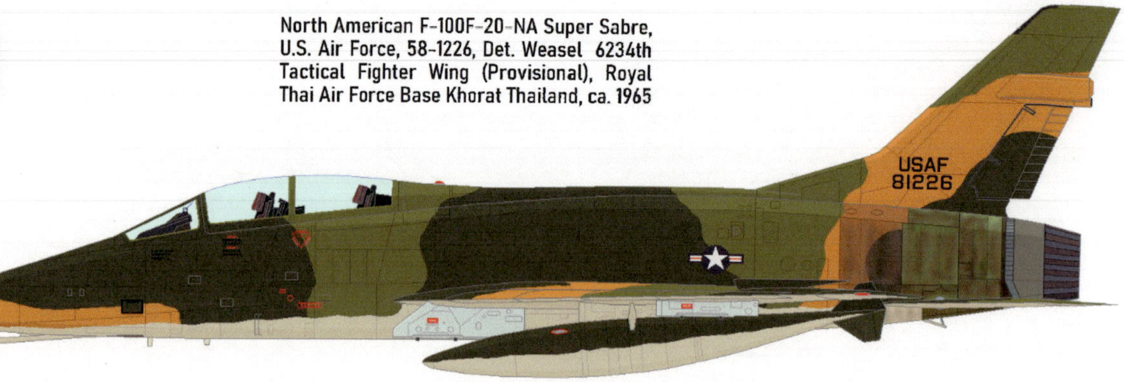

North American F-100D-90-NA Super Sabre, US Air Force, 56-3279, 113rd Tactical Fighter Squadron 181st Tactical Fighter Group, Hulman Field Indiana, ca. 1977

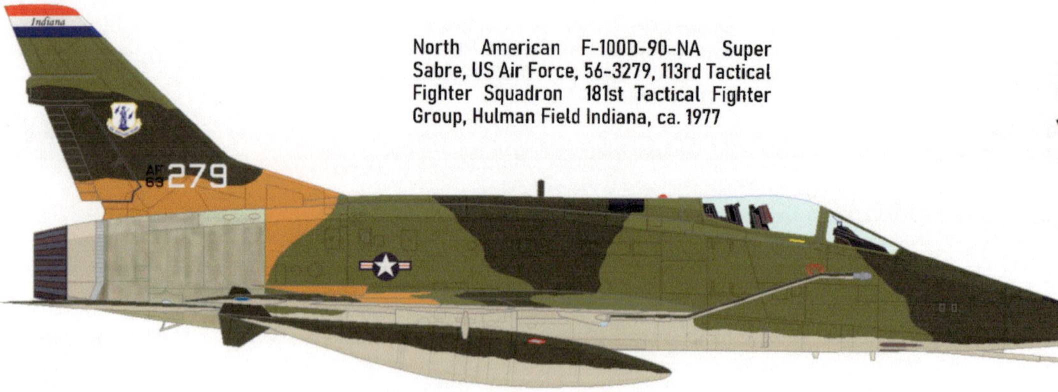

North American F-100F-15-NA Super Sabre, US Air Force, 56-3971, 4520 Combat Crew Training Squadron 57 Fighter Wepons Wing, Nellis AFB Nevada, ca. 1970

North American F-100D-65-NA Super Sabre, US Air Force, 56-2989, Tactical Fighter Squadron 4403 Tactical Fighter Wing, England AFB LA., ca. 1970

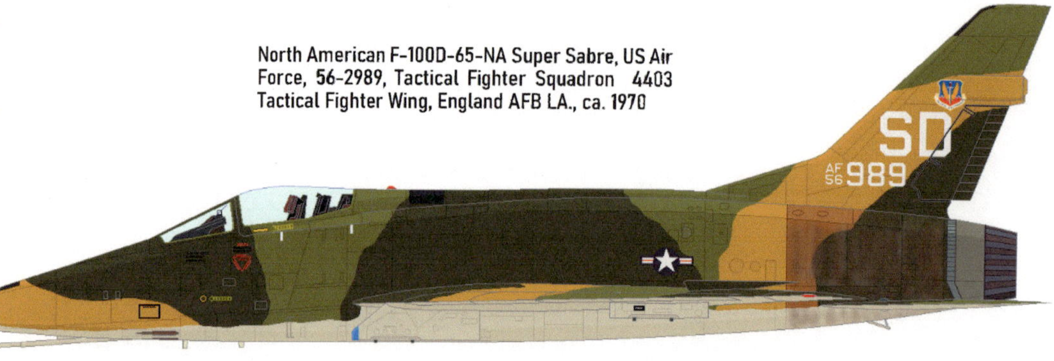

F-100D-30-NA Super Sabre (55-3710) 110th TFS 131st TFW. Robertson ANGB, Lambert Field, Saint Louis, Missouri 1978. Converted to QF-100D.

F-100D-90-NA Super Sabre (56-3298) 125th TFS 138th TFG. Tulsa International Airport, Oklahoma 1976. Converted to QF-100D.

F-100D-25-NA Super Sabre (55-3685) Arkansas ANG, 184th 'Flying Razorbacks', TFS 188th TFG, Fort Smith Municipal Airport, Arkansas 1976. Converted to QF-100D

F-100D-90-NA Super Sabre (56-3333) Georgia ANG, 128th TFS, Dobbins AFB, Marietta, Georgia 1975. Converted to a QF-100D.

F-100D Super Sabre of the 306th TFS at Tuy Hoa Air Base, Vietnam, in January 1970. (National

F-100D Super Sabres at Phan Rang Air Base, Vietnam, in February 1970. (National Archives)

F-100, 55-3752. It is seen here at Naval Air Station Atlanta in 1978. To MASDC October 6, 1978 as FE405. Converted to QF-100D drone no. 175. W/O 31 July 1985, shot down. (Acroterion)

F-100D, 56-3311, 306th TFW, in flight over South Vietnam in 1967. To MASDC August 10, 1979, as FE579. Converted to QF-100D drone no. 261. W/O September 8, 1989. Shot down. (USAF)

F-100D, 55-2881, of 31st TFS, 3rd TFW, from Bien Hoa Air Base, South Vietnam, in the late 1960s. This aircraft was retired to the MASDC on 18 April 1979 as FE0510. (USAF)

This F-100 pilot, of the 90th TFS, targeted the Vietcong in South Vietnam with napalm, in March 1966. (USAF Museum)

F-100D of 352nd TFS, dropping Napalm near Bien Hoa. (USAF)

F-100D of the 35th TFW, attacking target in Vietnam in September 1969. (USAF)

F-100D Super Sabre nicknamed 'Sizzlin Susie' of the 352nd TFS at Phan Rang Air Base, Vietnam, in August 1970. (National Archives)

F-100D-75-NA, 56-3183 of 118th TFS. To MASDC January 9, 1979 as FE446. Converted to QF-100D drone No. 166. W/O 20 Oct. 1986. Shot down by an AIM-7M missile. (USAF)

F-100D-85-NH, 56-3456, of the 355th TFS, at Tuy Hoa Air Base in 1969. To Turkish AF as 56-3456. (USAF)

F-100D-60-NA, 56-2943, 164th TFS. Later, 122 TFS NAS New Orleans ANG, LA. On Jun 3, 1976, the aircraft had an engine failure over Louisiana. The pilot ejected but suffered injuries. (USAF)

F-100F, 56-3868, 128th TFS. To MASDC January 3, 1979 as FE439. Converted to QF-100F drone no. 370. W/O 27 June 1989. Shot down by AIM-7M missile. (USAF)

F-100F-10-NA, 56-3882, of the 416th Tactical Fighter Squadron, 37th Tactical Fighter Wing at Phu Cat AB, South Vietnam. The 416th TFS operated F-100s from Phu Cat from April 15, 1967, to May 27, 1969. The F-100Fs were the 'Misty' fast FACs. This aircraft was retired to MASDC on Sep. 7, 1979, as FE0591. (USAF)

A North American OV-10A-30-NH Bronco, 67-14659, is flying in tandem with two F-100C Super Sabres of the 136th Tactical Fighter Squadron. The 136th TFS was a New York ANG unit assigned to the 31st Tactical Fighter Wing at Tuy Hoa Air Base, Vietnam, from June 14, 1968, to May 25, 1969. Tragically, the OV-10A was shot down on April 7, 1973, near Trapeang Veng, Cambodia, while on assignment with the 23rd Tactical Air Support Squadron of the 56th Special Operations Wing. (USAF)

F-100D-75-NA, 56-3198, of the 113th Tactical Fighter Squadron in Vietnam. Retired to MASDC on July 9, 1979, as FE0558. Later converted to a QF-100D target drone and destroyed. (USAF)

F-100F-10-NA, 56-3805, of the 127th Tactical Fighter Wing, Michigan Air National Guard, in the 1960s. This aircraft was sent to MASDC as FE0619 on Oct. 1, 1979. (USAF)

F-100D-65-NA, 56-3000, 182nd TFS. Currently on display outside at Texas ANG, Lackland AFB, Kelly Field, San Antonio, Texas. (USAF)

131st TFS, F-100D-25-NA, 55-3634, in 1975. It was with PACAF and equipped with the Lightweight Navigation System (LWNAVS). It was transferred to MASDC on September 7, 1979, as FE592. Converted to QF-100D drone No. 337. W/o on Feb. 9, 1990, after it was shot down. (USAF)

F-100 OPERATIONS IN VIETNAM

After the Gulf of Tonkin events, where the USS Maddox was attacked by three North Vietnamese torpedo boats, the USAF was pushed into the vortex of Vietnam. In early August 1964, in response to the Maddox attack, two squadrons of eighteen B-57s each quickly moved from Clark AB in the Philippines to Bien Hoa, in South-Central Vietnam, about 16 miles northeast of Ho Chi Minh City. The Air Force also rushed a composite air strike force to the area to be based at Da Nang, including one F-105 squadron, two F-100 squadrons, three C-130 transport squadrons, and six RF-101 reconnaissance aircraft, as part of a deployment known as 'One Buck.' Thus, the F-100s entered the Vietnam War in the summer of 1964. As for the B-57s at Bien Hoa, they were fortunate: they flew back to Clark Air Base on October 22. A week later, on November 1, 1964, Bien Hoa was attacked by mortar fire at 2 am, in a 30-minute barrage by the Vietcong. Of the remaining 18 B-57s at Bien Hoa AB, five were destroyed, and most of the rest were damaged. In addition, the attack damaged three A-1H Skyraiders, a Kaman HH-43 Huskie helicopter, and two Douglas C-47s.

Tragically, four US and two South Vietnamese personnel died in the attack, and 72 others were injured. The raid was a significant escalation in the Vietnam conflict. Before the incident in the Tonkin Gulf the previous August, the civilian leadership of Vietnam had consistently turned down requests from the USAF to deploy jets inside Vietnam. After the assault on the American destroyer Maddox, 36 B-57 Canberra bombers flew to Bien Hoa, and numerous squadrons of F-100 Super Sabres began flying in and out of Da Nang. Both actions followed the Tonkin attack. Then came the Bien Hoa attack. That tragic event increased the need for the Air Force to use jets in South Vietnam. Tactical fighter units made temporary deployments, rotating personnel out every ninety days and aircraft every six months. Despite this, using jets in armed conflict was still banned inside South Vietnam. The B-57 Canberras flew visual reconnaissance along routes leading into Saigon, while the F-100s flew in 'Operation Barrel Roll.' The USAF 2nd Air Division (later the Seventh Air Force) and US Navy Task Force 77 conducted a covert interdiction and close air support campaign in Laos between December 14, 1964, and March 29, 1973. 'Operation Barrel Roll' aimed to disrupt key Viet Cong supply trains and infiltration routes.

Air Force personnel clean up the debris after the attack on Bien Hoa AB on November 1, 1964. (USAF)

Bien Hoa AB was attacked by the Viet Cong not many hours later. (USAF)

1965

February 19, 1965, General Westmoreland used his emergency power to deploy jets for the first time in South Vietnam when B-57s from Bien Hoa attacked enemy troops east of Saigon. For many weeks, the B-57s continued to bomb the eastern Saigon region, supported by F-100s from Da Nang, which struck the enemy further north in II Corps. The Vietnamese government's strong prohibition on using jet planes inside South Vietnam began to weaken from that time. On March 9, the South Vietnamese government relaxed the constraints on jets, giving Westmoreland free rein to use jet aircraft as he saw fit.

THE WILD WEASELS

Due to the ferocity of North Vietnam's growing arsenal of anti-aircraft guns, automatic weapons, and SA-2 missiles pounding South Vietnam, the situation in Vietnam was dire in terms of heavy losses to aircraft, pilots, and other personnel. Furthermore, the enemy's weapons could not be targeted because of their proximity to the Hanoi-Haiphong area and the Chinese border. President Johnson would have exploded if any errant missiles had crossed the Chinese borderline, which would draw China into the war. Fortunately for the Air Force, a secret weapon arrived on the scene just in the nick of time. The newest secret weapon had a cartoon character-sounding name, but it was pretty effective; it was the 'Wild Weasel.'

In October 1965, four F-100F jets were designated 'Wild Weasel 1' by the Air Force after being modified to carry special equipment. The equipment included the APR-25 vector RHAW (Radar Homing and Warning) receiver, which detected S-band signals emitted by North Vietnamese SA-2 fire-control radars and early-warning/ground-controlled intercept radars. The equipment could also detect C-band signals (from the improved SA-2 missiles) and the X-band airborne intercept radar. The Wild Weasel F-100Fs also carried the APR-25 (WR-300) L-band warning receiver to detect missile guidance emissions. They also carried the IR-133 panoramic receiver, which could detect S-band signals at greater range than the APR-25. The aircraft were also equipped with a KA-60 panoramic camera and a dual-track tape recorder.

From October 11 to 18, 1965, the four modified F-100Fs underwent an accelerated test-and-training program at the USAF Tactical Air Warfare Center at Eglin AFB, Florida. The special Eglin radars (SADS-1 and SADS-2) simulated the North Vietnamese S- and C-band Fan Song radars. The test results showed that crews could detect the SADS radars from 253 miles at 10,000 feet and 46-57 miles at 150 feet. The four F-100Fs deployed to Korat, Thailand, on November 25, 1965, and fell under the operational control of the 388th Tactical Fighter Wing. They began flying missions on December 3, primarily to locate and destroy SA-2 installations. Threat warnings to strike planes and ELINT (Electronic Intelligence) gathering became secondary missions. In late 1965, the growing number of SA-2 installations prompted the Air Force to increase Wild Weasel 1 aircraft to seven and plan to deploy seven Wild Weasel III F-105F Thunderchiefs. This decision was made by the Air Council on December 30, 1965, and was approved by General McConnell on January 6, 1966. The F-105Fs were equipped with the same RHAW devices as the F-100F Wild Weasel 1 jets. The USAF modified three additional F-100Fs on January 24, 1966, and deployed them in SEA in late February of that year. The USAF completed modifications for three other F-105Fs on February 23 and deployed five aircraft to SEA in mid-April 1966. USAF planning also recommended deploying four Wild Weasel IV F-4Cs to Southeast Asia in mid-1966. However, this project did not materialize.

The Wild Weasel 1 missions were flown from Thailand in December 1965. The tasks were part of the 'Iron Hand' anti-SAM air campaign (Iron Hand was initially organized in early August 1965 as a special F-105 ground alert force designed to detect and suppress enemy surface-to-air-missile sites. However, the operation made little progress. Consequently, the role of the Iron Hand F-105s was changed, and they were tasked with searching for SAM locations over the border in North Vietnam.

An electronic warfare officer in the back seat operated the F-100F radar homing and warning equipment. The F-100 Wild Weasels used 'Search and Destroy' tactics, which the Air Force developed for the Wild Weasel 1 missions. The strike force included one Wild Weasel F-100F carrying two rockets and 20-mm ammunition, typically leading three F-105s, each armed with two rocket pods and three 500- or 700-pound bombs. As the hunter aircraft, the F-100F attempted to home in on the SA-2 radar emissions by flying a planned track between 4,500 and 15,000 feet. At the same time, the electronic warfare officer monitored the special RHAW equipment. During the search, the crew used direct homing or a circular approach. The F-105 killer aircraft trailed the Wild Weasel 1 by about 15 seconds. After acquiring the target electronically, the F-100F pilot would eyeball it and mark it with rockets. On seeing the F-100F rockets striking the mark, the attacking F-105s would then drop their ordnance on the missile sites.

As mentioned previously, the F-100F Wild Weasel 1 aircraft also served as threat-warning platforms, accompanying strike aircraft and alerting them to stay above ground fire on their way to the target zone. EB-66 Destroyers flew ELINT missions, collecting information for planning future search-and-destroy missions. Lone Wild Weasel aircraft would occasionally fly ELINT missions along the North Vietnamese border to gather data and familiarize crews with the signal environment. Starting on April 3, 1966, the F-100Fs and F-105s were equipped with the AGM-45 Shrike anti-radiation missile, designed to home in on hostile anti-aircraft radar. That allowed the Wild Weasel planes to attack the 'Fan Song' fire control radars. The lead aircraft, usually the F-100F, delivered the Shrike missile in a dive, while the rest of the Wild Weasels maneuvered to home in on the radar and destroy it.

Between late November and December 23, 1965, the F-100Fs helped develop operational tactics and gain experience on Iron Hand missions. However, poor weather in the target areas canceled many sorties and hampered others. As a result, all 'Iron Hand' attacks were halted during the bombing truce from December 24, 1965, to January 30, 1966. Instead, the Wild Weasel aircraft flew 49 ELINT missions over Laos and the Gulf of Tonkin. Again, poor weather from January 31 to late March hampered Iron Hand operations. However, the Wild Weasels destroyed two SA-2 installations during that period. Then, as the weather improved significantly in April and May, the Wild Weasels hit five SAM installations, inflicting heavy damage on three sites.

From 65 to May 15, 1966, 394 Wild Weasel 1 sorties were planned, but lousy weather canceled 247, leaving only 147 Iron Hand sorties. Of that total, the Wild Weasels attacked only 13 sites.

The Wild Weasels saw SA-2 batteries in only five cases because enemy camouflage made targets hard to eyeball. After July 23, 1966, the Wild Weasels destroyed or damaged ten SA-2 sites. Shrike missiles also damaged six SA-2 vehicles and destroyed or damaged five AAA stations. The strikes also caused minor damage to 11 AAA radars. Meanwhile, two F-100Fs and two F-105Fs were lost in combat, while one F-100F suffered engine failure. Ultimately, the attacks caused minimal damage to the SAM sites. Several factors contributed to this: the Wild Weasel force could not visually acquire targets because the SAM sites were heavily camouflaged. Moreover, the weather was not suitable for any attacks. The 4,500-foot altitude restriction imposed on Wild Weasel aircraft for safety reasons also played a part, as the pilots could not see the targets from that height.

Furthermore, the lack of ranging equipment forced the F-100Fs to fly in close to the SA-2 site and then pull up rapidly to capture it visually. That tactic was dangerous as it made the aircraft vulnerable to ground fire. In addition, target restrictions prevented USAF aircraft from attacking some batteries. For political reasons, the Rules of Engagement (ROE) imposed on US forces in SEA caused this. Eventually, the Wild Weasel aircraft proved to be a negligible response to the missile threat. However, they did assist the USAF's campaign somewhat, as they made the operations of the North Vietnamese missile sites more problematic due to the perceived threat of the Wild Weasels.

'The Wild Weasels': These pioneers made history. They ushered in the Wild Weasel and, over time, developed effective tactics that destroyed nine SA-2 sites and suppressed enemy SAMs for strike forces. The pioneers—Wild Weasel Detachment, 6234th Tactical Fighter Wing, Korat, Thailand. The first Wild Weasel aircrews (front, left to right) are: Capt. Walt Lifsey, Capt. Sandy Sandelius, Capt. Ed White, Maj. Garry Willard, Capt. Jack Donovan, Capt. Allen Lamb, Capt. John Pitchford, Capt. Maury Fricke, unknown, and Maj. Bob Swartz (not pictured are Capt. Les Lindenmuth, Capt. Donald Madden, and Capt. Robert Trier). (USAF)

Wild Weasel commander Maj. Garry Willard at Korat, Thailand. He later returned to the U.S. to head the Wild Weasel school at Nellis AFB. Willard later retired from the USAF as a brigadier general. (USAF)

"Willy" Wild Weasel 1. This is the official logo of the Wild Weasel crew (USAF)

Pictured here is most of the second group of F-100F Wild Weasels at Korat, Thailand, in early 1966

One of the F-100Fs was converted to a Wild Weasel aircraft at Eglin AFB, Florida, in 1965. (USAF Museum)

OPERATION BLACK VIRGIN MOUNTAIN

Intelligence identified multiple enemy base camps in an area where Viet Cong preparations were underway in early 1965. It was in the notorious area, War Zone C, in the Tay Ninh Forest, sixty miles northwest of Saigon and North of Tay Ninh City, near the Cambodian border. Since World War II, this zone, named by the French, had defied all attempts to penetrate it. It was virtually impenetrable.

The base camp served as the Vietcong's headquarters. A signal school was established near the base camp to train local recruits and North Vietnamese subversives. The enclave contained over 100 buildings. Some structures were built of wood, but most were fashioned from jungle materials. The structures were divided into four complexes with residential quarters, mess halls, schools, arms-production units, and anti-aircraft positions. Several shelters and bunkers dotted the complex, all linked by narrow ditches. The entire enclave sat atop a complex network of tunnels. The jungle canopy rendered the area below the foliage completely undetectable from the air and inaccessible from the surface. The complex was like Fort Knox; no one could get in.

In mid-April, Westmoreland planned a massive air campaign against the installation, nicknamed after the neighboring 'Black Virgin' Mountain. Westmoreland's initial plan called for an all-out air campaign using USAF planes from both South Vietnam and Thailand, as well as, for the first time, US Navy and Marine aircraft. However, under the rules of engagement, USAF aircraft from Thailand could not fly directly to the target zone; instead, they had to land at a South Vietnamese airstrip. The South Vietnamese were also ready to join the offensive, but American authorities informed them that some of their aircraft would need to remain in South Vietnam in case of a threat from the Vietcong, NVA, or both.

Westmoreland had hoped to suspend all other air missions in South Vietnam and Laos, as well as 'Rolling Thunder' in the North, for this one-day operation. General Moore of the 2nd Air Division would organize, coordinate, and lead the entire air attack.

Backup elements, including Air Force search-and-rescue helicopters, an ordnance demolition squad, and an Army medical evacuation helicopter, arrived at Tay Ninh Airport on April 15. The first bombs were dropped on the facility at seven a.m., and the bombardment proceeded nonstop for 12 hours. Air Force B-57s and A-1E Skyraiders from Bien Hoa, and F-100 Super Sabres from Da Nang, flew an astounding 206 sorties. The South Vietnamese also provided 54 A-1H Skyraider sorties, and the USMC kicked in with 26 F-4 B sorties. US Navy assault aircraft from the carriers USS Midway and USS Coral Sea brought the total to 417.

The 2nd Air Division coordinated and directed the strikes through its airborne FACs, who used their Bird Dogs to mark targets for the attacking aircraft. Hovering above the action, two U-10 Helio Couriers alternated with the Bird Dogs in orchestrating the operation. By mid-afternoon, the 1.9-mile-by-3.7-mile area was engulfed in smoke from the bomb strikes and grass fires. One explosion sent a white cloud billowing 3,000 feet into the air. As the day wore on, large sections of tree cover burned off, exposing hidden buildings and revealing entrances to underground installations. By the time the attacks ended at seven in the evening, 816 tons of bombs had been dropped on the enclave. It was total overkill. Fort Knox of the Jungle was no more.

For two days afterward, 1,200 Army of the Republic of Vietnam (ARVN) soldiers, flown into the area by helicopter, fanned out in search of the Vietcong and to examine the results of the raids. Air Force and Vietnamese A-1s and US Marine F-4s provided air cover and close air support for the searchers. While the troops saw no enemy, they uncovered ample evidence of their recent departure. Sixty buildings had been destroyed from the air, and the South Vietnamese captured switchboards, radios, communication wires, and various ordnance. The F-100s that participated in the raids completed the task as required. Finally, they flew back to Da Nang for a well-earned break.

1966

COMBAT SKYSPOT

As operational requirements increased, the F-100 underwent modifications to improve combat performance, some of which were nearly 14 years old. The aircraft's weapon release and firing mechanisms were overhauled, and new weapons were installed. The F-100 also received a new bombing system nicknamed 'Combat Skyspot,' a modification program for ground-directed bombing (GDB) operations. It was installed on the F-100s in SEA in April 1966. In the early days of the Vietnam War, the Air Force's aircraft could not deliver ordnance with reasonable accuracy under conditions other than good weather and daylight. The Strategic Air Command (SAC) had the answer. SAC used a simple method to assess its aircrews' bombing skills.

No 'real' bombs were involved. A modified gun-laying radar tracked the aircraft, and operators plotted the bomber's track on paper. The bomber crew would indicate when the simulated bomb would be released. The bomb drop point was normally logged by an electronic tone that activated a few seconds before the simulated release and then switched off at release. This action created a break in the aircraft's recorded track. If the system received correct altitude, ground speed, airspeed, and wind data, it could check a ballistics database for a specific bomb type and apply all external data to the recorded ground track to produce an accurate score. SAC reversed this scoring procedure and fine-tuned many elements of the system. As a result, SAC created a precise ground-guided bombing system. This new system worked with most bomb types. The (TSQ-77) upgrade, which equipped each aircraft with Motorola's new SST-181X band radar transmitter, gave all F-100s flying at night or in poor weather ground-directed bombing capability. Interestingly, between 1966 and the end of combat operations in 1973, 'Combat Skyspot' teams successfully directed thousands of B-52 strikes and tactical bombing missions in SEA.

Most aircraft dropped their bombs between 10,000 and 20,000 feet. Skyspot quickly gained popularity, and before long, radar handled one-fourth of tactical strikes. It wasn't until July that the B-52s started using the technique, and by year's end it became the Buffs' primary means of bombing.

The F-100s also carried out horizontal bombardments out of sight of the ground during 'Combat Skyspot' missions. They conducted these missions with a radar on the ground and a suitable receiver in the F-100D cockpit. The pilots dropped the bombs when two light signals crossed on their receiver. For a time, 'Combat Skyspot' equipped F-100Ds also acted as guides for other fighter-bombers, with the Hun pilot giving those following him the signal of when to drop their loads.

Unfortunately, 'Combat Skyspot' was not the hoped-for single solution to the problems of night and bad-weather bombing, and in terms of sortie rates, it was not the most effective system. Its radars could reach out only 50 miles. Although special beacons on fighter planes could amplify the signal and extend the range to 200 miles, such beacons were scarce in 1966, and only the F-100s at Bien Hoa carried them. Because a Super Sabre carried only one-fourth the bomb load of a B-57, it took three or four F-100 sorties to do what one Canberra raid could. Like all radar, the Skyspot beam could not pass through obstacles and needed an unimpeded path to the aircraft, limiting how low the fighter planes could fly. The radar could control only one flight at a time against one target, far less effective than a forward air controller, who often orchestrated several flights over a target at once.

Skyspot, despite its flaws, gave the Air Force a much-improved, albeit restricted, capability to strike the enemy at night and in adverse weather. The system was modified to allow phased adoption. First, the regulations that had previously confined air strikes to those under the authority of an FAC were altered to enable strikes under the control of 'Combat Skyspot.' However, the new system remained a backup and was used only when FACs and flares were unavailable.

1966-1967

'MISTY' AND THE TALLY HO CAMPAIGN

Tally-Ho was an air interdiction operation against the VC and North Vietnamese Army (NVA), which took place in the panhandle of North Vietnam, along the northern border of South Vietnam. The campaign commenced in July 1966 following the North Vietnamese invasion of Quang Tri Province. The FAC Flying 0-1 Bird Dogs out of Tiger Hound forward airstrips (airstrips located inside Laos, where Air Force, Navy, and Marine aircraft concentrated attacks. It was a 125-mile section of the southeastern corner of Laos. Operations within the sector were nicknamed 'Tiger Hound'. The Tally Ho controllers worked in pairs but failed to match 'Operation Cricket's successes. ('Operation Cricket' began in January 1966. It included the territory between southern Tiger Hound and Nape Pass to the North. It was concentrated on the North Vietnamese border, where the trail crossed into Laos. The USAF viewed Cricket as an extension of the war in South Vietnam, and MACV oversaw actions. The 'Cricket' operation used Forward Air Controllers (FACs) and involved the use of slow movers, 0-1 Bird Dog, and A-1E Skyraider aircraft. FACs directed strikes by attack aircraft against promising enemy positions. Rapid assaults on targets were possible because of instantaneous communication between airborne FACs and the Royal Laotian Air Force (RLAF) headquarters. Tally Ho was a different situation. The USAF had less success because stout enemy defenses in southern North Vietnam's coastal areas forced FACs to fly too high for effective Visual Reconnaissance (VR). When the FAC aircraft were on their target-marking pass, enemy gunners massed their fire. This forced the Bird Dogs to rise to a minimum altitude of 2,500 feet to avoid being hit. From that height, spotting the Vietcong and NVA positions was not easy. As a result, FAC accuracy dropped sharply. By the end of 1966, the A-1 Skyraider, an excellent loitering attack aircraft, replaced the 0-1 Bird Dogs, which moved to less dangerous mountain areas to conduct FAC operations. However, the A-1 Skyraiders, in turn, gave way to the F-100 FACs, nicknamed 'Misty.' Misty would show the way.

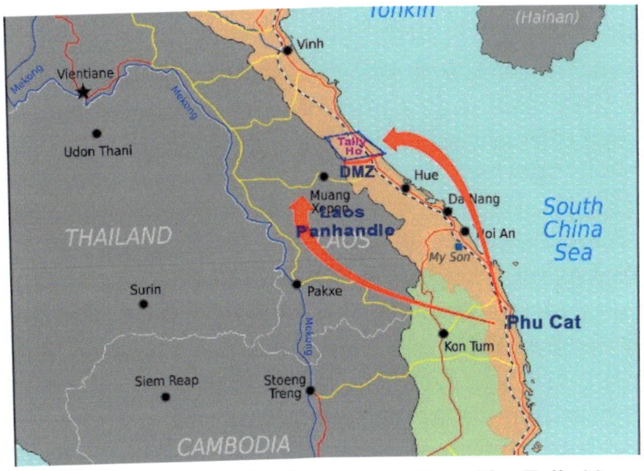

This is a map of the Tally Ho operation. The Tally Ho targets were just across the border in southern North Vietnam. (Graphics by author, map by User: (WT-shared) Burmesedays)

In the panhandle of North Vietnam and southern Laos, stiffening enemy defenses made it very risky for the slow-moving A-1 Skyraider FACs to conduct their mission. As a result, on May 17, 1967, the Seventh Air Force Commander authorized 'Operation Commando Sabre,' a test of the F-100F's capacity to undertake armed reconnaissance and FAC missions. The F-100F had many advantages, including high speed and rapid maneuverability, which would enable it to survive in high-threat environments. Moreover, the Hun was well-suited for armed reconnaissance because it carried two LAU-59 B/A rocket launchers for target marking and dependable M39 20-mm guns. In addition, external fuel tanks (and in-flight refueling) would lengthen the duration of the FAC operations.

On June 28, 1967, Detachment 1 of the 416th Tactical Fighter Squadron, 37th Tactical Fighter Wing, was assigned to the Super Saber mission dubbed 'Misty,' the radio callsign used by the F-100F Fast Forward Air Controllers (Fast FACs) during the Vietnam War. The Seventh Air Force gave the detachment permissive parameters and the liberty to experiment.

Misty FAC volunteers needed to be flight leaders with at least 100 strike missions in SEA and 1,000 flying hours. One hundred and twenty days, or 75 missions, whichever came sooner, was the initial duty tour. Twelve of the initial 16 recruits lacked FAC expertise, having come from F-100 units in South Vietnam. Four fighter-qualified controllers from the 504th Tactical Air Support Group instructed the other four. The newly formed unit acquired F-100F fighter jets from local fighter units and began training the recruits.

The Misty pilots were familiar with conditions in southern North Vietnam. They knew almost every supply route, trail, and road that existed. Once in the air, 'Misty' began hunting for targets for the attack planes. The 'Misty' F-100Fs were on the lookout for apparent targets such as anti-aircraft gun installations, supply trucks, and structures, as well as anything unusual, such as a square bush or a circular formation. There were other times when a Misty pilot noticed water on one side of a river crossing and knew the vehicles were on that side of the river. After locating the targets, Misty pilots called in strike pilots and guided the strike aircraft in and out of enemy territory. Misty pilots briefed strike pilots on target locations, defenses, and the best escape routes. They used smoke rockets to mark targets, making the attack aircraft' jobs easier.

The attacking aircraft formed a box pattern around the target, each rolling in one after the other, spaced out so their guns stayed on the target. At the same time, the Misty F-100F in front of them pulled away. The attacking pilots preferred a 'wheel' pattern, letting them roll in from various headings to confuse the gunner's tracking. However, with friendly forces nearby, accuracy was the name of the game for the attacking squad, even if it meant losing a round or two, which was better than causing a friendly-fire incident. The F-100F FAC would watch the attack and frequently inform the attacking pilots what heading to take, using a basic code.

The Misty FACs performed better than expected. They flew 82 sorties in July 1967 alone, ordering 126 attacks on targets that their own VR spotted. In the same month, they discovered 150 truck stops, bridges, shallow points in a river, and other locations suitable for road interdiction. The Fast FACs operated in hostile territory where slower FAC planes, like the Bird Dog and the A-1 Skyraiders, could not fly. Furthermore, as the 'Misty' FACs gained more experience, their ability to find enemy target positions became more and more remarkable.

Two months later, in September and October 1967, the 'Commando Sabre' FACs experienced their first large-scale combat. Throughout the summer, North Vietnamese artillery barrages from across the DMZ pounded positions at Con Thien, Gia Linh, Camp Carroll, and Dong Ha. Finally, on September 12, 'Operation Neutralize' (a 6-week air-and-ground assault) began to counter the strikes. Air attacks were commanded by 20 0-2A (Cessna Skymaster) FACs south of the DMZ, while the Mistys took care of the North Vietnam territory over the border.

The Misty F-100Fs were bombarded with Air Tasking Order (ATO) 'Frag' sorties ('frags' are details of the squadron's missions for the day). From the outset, the 'Mistys' focused on priority targets provided by the Seventh Air Force and uncovered additional targets. Almost all targets were artillery positions, so small they needed a direct strike to be destroyed. Since the pilot was not authorized to fly below 4,500 feet and had to jink (perform evasive maneuvers) repeatedly, the FAC had difficulty pinpointing and marking these AA locations. Despite these shortcomings, figures from 'Operation Neutralize' showed that attacks by fighters under Misty's command were twice as successful as those carried out on their own. The 'Commando Sabre' operation particularly impressed General Momyer, the commander of the Seventh Air Force, who would in the future use the 'Misty' in other challenging operations, which needed fast mover FAC aircraft.

While finding targets, the F-100F pilot and pilot-observer learned to work well together. During VR, the rear-seat observer conducted most of the searching, mapping, and documenting of sighted targets. He also photographed immovable and mobile targets, such as parked trucks, truck parks, and bridges, using a hand-held camera. After both crew members confirmed the prospective targets, the rear-seater requested attack aircraft from the 'Airborne Battlefield Command and Control Center' (ABCCC). The pilot handled the rendezvous, target marking, and strike control.

Meanwhile, the pilot-observer monitored everything else. One guy could do the job, but two made it much easier. The Misty F-100Fs played a significant part in the Vietnam conflict.

In 1967, when the 'Commando Sabre' program was first initiated, the Air Force had already begun preparations to phase out the F-100s in 1970. As a result, in early 1968, the Seventh Air Force began formulating plans for the F-4 Phantom's fast-FAC program. Several issues had plagued Misty's activities. The F-100F lacked the radar to detect impending SAM or AA strikes and the ECM pods needed to defend against them. Without the radar to warn the Misty pilots of an attack, the Misty FAC pilot would not become aware of one until it was reported to him by either the ABCCC ground radar or another aircraft. Even if the pilot was given advanced warning, he could not direct strikes on the SAM/AA sites unless he could identify them with his naked eye. In addition, the F-100F's lack of power made it especially vulnerable to ground fire during evasive maneuvers. Let's face it, the Sabre was built to be fast, but there was not a lot of emphasis on maneuverability, and some did get hit by ground fire. However, the biggest obstacle was that they never had enough aircraft to meet strike-operation requirements. Despite these limitations, the 'Commando Sabre' experiment demonstrated the value of the Fast FAC and laid the foundation for the fundamental strategies used in F-4 FAC operations.

The more powerful F-4 Phantom, available in greater numbers, was considered a replacement for the F-100F. However, trials revealed issues with the F-4. For instance, the visibility from the Phantom's back seat was poor. It used more gasoline at low altitudes than the F-100 and required more refueling. Moreover, its turn radius was even greater than the Super Sabre's, and staying close to the target increased its fuel consumption even more, limiting the time the F-4 could remain in the conflict zone. The tests found the F-4 Phantom acceptable but not significantly better than the F-100Fs. However, as planned, the Air Force withdrew the F-100Fs from FAC operations in 1970. When the final 'Misty' mission took to the skies on May 14, 1970, the F-4 program was already well underway. Over those three years, from 1967 to 1970, the pilots of the 'Misty' Huns pioneered the 'fast mover' FAC role, paving the way for the F-4s to take over. The F-100 Misty pilots all put their lives on the line daily, and some were tragically lost. Of the 157 Misty pilots who served as FACs, 34 were shot down, eight were killed, and four became prisoners of war. The 'Misty' pilots did an extremely perilous job, and they did it very well. They are true heroes and will be remembered for their feats in a hostile environment.

F-100F-20-NA, 56-1222, of the 612th TFS, fires a smoke rocket towards a target over Vietnam, circa 1968. Detachment 1 of the 612th TFS (tail code 'HS') was assigned to the 37th TFW, at Phù Cát Air Base, Vietnam, from June 1967 to April 1969. The rest of the squadron was with the 35th TFW at Phan Rang (tail code 'VS'). This F-100F is a fast Forward Air Control (FAC) 'Misty'. (National Archives)

This is the plane, and these are the pilots who flew it into the dragon's mouth with minimal firepower, using smoke rockets to guide the attacking squad to its targets. They never flinched or turned away from the objective. They did their job as professionals, and because of that, they stand tall in everyone's eyes as heroes. This reunion, held in October 2014, was hosted by the Pacific Aviation Museum in Hawaii. (Photo: Pacific Aviation Museum)

(Photo: USAF MUSEUM)

This is a photo of F-100F 'Misty' Fast FAC pilots Major John F. Overlock and Major Michael D. McElhanon, who served with the 309th Tactical Fighter Squadron at Phu Cat, South Vietnam. 'Misty' FAC operations over North Vietnam were their most common missions. Overlock was classified as a co-pilot, while McElhanon was rated as a pilot.

On August 16, 1968, McElhanon and Overlock were on an early mission and had already refueled once. They radioed Airborne Control that they were heading to the Gulf of Tonkin for a second refueling with a tanker. After that, Airborne Control had no further contact with Overlock and McElhanon. They were not reported missing until fifty minutes later, when several fighter jets tried to locate them to acquire a fix on their intended target. However, the two pilots were never seen or heard from again. Tragically, their aircraft was believed to have crashed near the city of Dong Hoi in Quang Binh province, North Vietnam. While they were missing, Michael O. McElhanon (34) was promoted to Colonel, and John F. Overlock (32) to Lieutenant Colonel. To this day, their fate remains unknown.

F-100 PRODUCTION LIST

SERIAL NUMBERS	MODELS	UNITS PRODUCED	CONSTRUCTION NUMBERS
52-5754/5755	YF-100A-NA	2	180-1/2
5756/5765	F-100A-1-NA	10	192-1/10
5766/5778	F-100A-5-NA	13	192-11/23
53-1529/1568	F-100A-10-NA	40	192-24/63
1569/1608	F-100A-15-NA	40	192-64/103
1609/1708	F-100A-20-NA	100	192-104/203
1709/1778	F-100C-1-NA	70	214-1/70
54-1740/1769	F-100C-1-NA	30	217-1/30
1770/1814	F-100C-5-NA	45	217-31/75
1815/1859	F-100C-15-NA	45	217-76/120
1860/1970	F-100C-20-NA	111	217-121/231
1971/2120	F-100C-25-NA	150	217-232/381
2121/2132	F-100D-1-NA	12	223-1/12
2133/2151	F-100D-5-NA	19	223-13/31
2152/2221	F-100D-10-NA	70	223-32/101
2222/2303	F-100D-15-NA	82	223-102/183
55-2709/2733	F-100C-10-NH	25	222-1/25
2734/2743	F-100D-35-NH	10	224-1/10
2744/2783	F-100D-40-NH	40	224-11/50
2784/2863	F-100D-45-NH	80	224-51/130
2864/2908	F-100D-50-NH	45	224-131/175
2909/2954	F-100D-55-NH	46	224-176/221
3502/3601	F-100D-20-NA	100	223-184/283
3602/3701	F-100D-25-NA	100	223-284/383
3702/3814	F-100D-30-NA	113	223-384/496
55-5118/5120	YF-107A-NA	3	212-1/3
5121/5126	YF-107A-NA	6	Not Produced
56-2903/2962	F-100D-60-NA	60	235-1/60
2963/3022	F-100D-65-NA	60	235-61/120
3023/3142	F-100D-70-NA	120	235-121/240
3143/3198	F-100D-75-NA	56	235-241/296
3199/3346	F-100D-90-NA	148	235-297/444
3347/3350	Canceled	------------	-----------
3351/3378	F-100D-80-NH	28	245-1/28
3379/3463	F-100D-85-NH	85	245-29/113
3725/3739	F-100F-1-NA	15	243-1/15
3740/3769	F-100F-5-NA	30	243-16/45
3770/3919	F-100F-10-NA	150	243-46/195
3920/4019	F-100F-15-NA	100	243-196/295
58-1205/1233	F-100F-20-NA	29	255-1/29
6975/6983	F-100F-15-NA	9	261-1/9 (MDAP)
59-2558/2563	F-100F-15-NA	6	262-1/6 (MDAP)

Creative Commons Credits

BIBLIOGRAPHY

American Combat Planes - Wagner, Ray 1982
The North American F-100 Super Sabre, Profile Publications 1965
Encyclopedia Aircraft & Missile Systems - Knaack, Marcelle Size 1978
US Army and Air Force Fighters 1916-1961 Harleyford Publications Limited 1961
Combat Aircraft of The World - Swanborough, Gordon 1962
The American Fighter - Angelucci, Enzo, and Bowers, Peter 1987
Rain of Fire - The Vietnam Experience - Air War, 1969-1973 Morocco, John 1985
A Pictorial History of the Vietnam War - Newcomb, Richard F. 1987
A War Too Long - The USAF in Southeast Asia 1961-1975 - Schlight, John 1996
An Illustrated Guide to the Air War Over Vietnam - Nalty, Bernard C. 1981
Air Tactics Against NVN Air-Ground Defenses December 1966 - Project Checo 1969
Air Power In Three Wars - General William W. Momyer, 1978
Tactics and Techniques of Close Air Support Operations, 1961-1973 Lt. Col. Ralph A. Rowley, 1976
Interdiction in Southern Laos, 1960-1968 - Van Staaveren, Jacob, 1993

Made in the USA
Las Vegas, NV
03 November 2026

57423047R00048